Next-Gen AI Robotics

Humanoids Transforming the Factory Floor

by
Charlie Addison

Next-Gen AI Robotics

Humanoids Transforming the Factory Floor

Contents

Introduction

The world of industrial automation is at the cusp of a revolution, driven by the advances in artificial intelligence (AI) and robotics. Central to this transformation is the emergence of AI-powered humanoid robots, machines that mimic human capabilities and behavior more closely than ever before. Designed to operate in environments traditionally dominated by human workers, these robots promise to bring about significant changes in manufacturing processes, operational efficiency, and the global labor market.

Advancements in robotics have come a long way. From the early days of mechanical arms that could perform repetitive tasks on assembly lines to today's sophisticated humanoid robots, the journey has been one of continuous evolution. These advancements are not just technological; they are also about enhancing our capacity to solve complex problems and meet the growing demands of a highly competitive global market.

Humanoid robots equipped with AI are unique in their ability to understand and respond to their surroundings. Combining advanced sensors, machine learning algorithms, and real-time data processing capabilities, these robots can perform a myriad of tasks with unparalleled precision and adaptability. They are not just tools; they are co-workers capable of interacting with human colleagues, learning from them, and even enhancing their productivity.

At the heart of this transformation are several driving forces. The pressing need for increased efficiency, the relentless quest for higher quality, and the undeniable benefits of reducing human error and workplace accidents make the case for AI-powered humanoids compelling. These robots are designed to work alongside human workers, complementing their skills and augmenting their abilities. This collaboration is key to achieving the seamless integration of robotics into the industrial landscape.

Industries that have traditionally been the pioneers in adopting robotics, such as the automotive sector, are demonstrating how AI-powered humanoid robots can reshape production lines. By accelerating production speeds and enhancing the quality of output, these robots are setting new benchmarks for what can be achieved in manufacturing. As industries across the board begin to recognize the potential of these advanced machines, the ripple effect will be felt worldwide.

The introduction of AI-powered humanoid robots is not without its challenges. There are technical limitations to overcome, ethical considerations to address, and the inevitable impact on the workforce to manage. However, history has shown that technological advancements often usher in new opportunities, redefining roles, and creating new career paths. As businesses and educational institutions adapt to this new reality, they will play pivotal roles in shaping the future workforce.

Equally important is the need to establish regulatory and safety standards that ensure these robots are integrated responsibly and safely into industrial settings. As stakeholders work together to develop these standards, they will help foster an environment of trust and acceptance among the public and the workforce. A collaborative approach will be essential in addressing the broader ethical and societal implications of deploying AI-powered robots.

This book aims to provide a comprehensive exploration of the myriad ways in which AI-powered humanoid robots are transforming industrial environments. We will delve into the core technologies driving these innovations, the key components that make up modern humanoid robots, and the software architectures that underpin their operation. We will also examine the economic impacts, regulatory frameworks, and the future trends set to shape this exciting field.

The chapters that follow will offer detailed case studies, highlighting successes and lessons learned from early adopters. We will explore the economic benefits and challenges of integrating these robots into various industries. Moreover, we will look at how companies are training their workforces to collaborate effectively with their robotic counterparts, fostering a symbiotic relationship between man and machine.

As we navigate this discussion, it is essential to keep in mind the global context. Robotics and AI are not confined by geographical boundaries; they are global phenomena with far-reaching implications. By examining case studies from different regions and understanding international collaborations, we can gain insights into how diverse cultural and regulatory landscapes are shaping the adoption of humanoid robots.

Looking ahead, the potential for humanoid robots extends beyond current manufacturing environments. Emerging trends and predictions point to their expansion into new industries and sectors, each with its unique set of challenges and opportunities. From healthcare to logistics, these robots have the potential to revolutionize numerous fields, making our exploration even more pertinent.

In addition to technological and economic considerations, it is crucial to address the human aspect of this transformation. Humanizing humanoid robots, building emotional intelligence, and enhancing their communication and interaction capabilities will be vital in ensuring that these machines are not only efficient but also relatable and user-friendly. Designing robots with ergonomics and human factors engineering in mind will play a significant role in their acceptance and effectiveness.

As we embark on this journey, it is clear that the future of industrial automation is bright and filled with opportunities. By harnessing the power of AI and robotics, we can achieve levels of efficiency and innovation previously thought impossible. This book is an invitation to explore this exciting frontier, to understand the technologies and trends that are shaping the future, and to be inspired by the endless possibilities that lie ahead.

Welcome to the new era of industrial automation, where AI-powered humanoid robots are not just a vision of the future but a present-day reality, poised to redefine the landscape of manufacturing and beyond. Together, let's discover how these remarkable machines are enhancing our capabilities, transforming our industries, and setting the stage for a future where humans and robots work hand-in-hand to achieve unparalleled success.

Chapter 1:
The Rise of Humanoids in
Industrial Settings

The advent of humanoid robots in industrial settings marks a transformative era in manufacturing, reflecting unprecedented advancements in artificial intelligence and robotics. These robots, designed to emulate human capabilities, are revolutionizing factory floors by enhancing operational efficiency and adaptability. Gone are the days when robots were limited to repetitive and monotonous tasks; today's humanoids integrate seamlessly into complex production lines. By utilizing sophisticated AI algorithms, they can perform intricate duties, learn from their environment, and work alongside human counterparts with remarkable dexterity. This integration not only boosts production speed and accuracy but also opens new avenues for innovation in the industrial sector. As we delve into this journey, we uncover how these intelligent machines are not just tools but pivotal partners in reshaping the future of global manufacturing.

The Evolution of Robotics in Manufacturing

The integration of robotics into manufacturing has been nothing short of transformative. From the primitive and rudimentary machines of the early 20th century to today's advanced AI-powered humanoids, the journey of robotics in the industrial sector is a testament to human ingenuity and relentless pursuit of efficiency. As we set the stage to discuss the rise of humanoids, it's critical to understand how robotics has evolved and laid the groundwork for this revolution.

Initially, industrial robots were simple, repetitive machines designed to perform specific tasks. They could be relied upon for operations such as welding, assembly, and painting – tasks that were deemed dangerous or monotonous for human operators. These early robots were programmed with fixed sequences of instructions, lacking the flexibility and adaptability that characterizes modern machines. Nonetheless, their introduction marked a pivotal shift, heralding an era where machinery began to take on roles traditionally held by humans.

In the 1960s and 1970s, companies like General Motors started integrating the first generation of robots into their production lines. These robots, such as the Unimate, were groundbreaking at the time. Unimate was capable of performing tasks like die casting and welding with precision far superior to that of human workers. However, they operated within a strict framework of programming and had limited sensory feedback, making them useful only in highly controlled environments.

The 1980s saw significant advancements in computer technology, which began to influence the development of industrial robots. Microprocessors provided more computing power, enabling robotics to become more sophisticated in their operations. Robotics started incorporating feedback systems using sensors, allowing them to perform more complex tasks with improved accuracy. During this period, Japan emerged as a leader in robotics, with companies like FANUC and Kawasaki revolutionizing manufacturing with their robotic arms.

As we approached the new millennium, the emphasis shifted towards integrating robotics with emerging computer technologies like artificial intelligence (AI) and machine learning. AI started to play a crucial role in enhancing the capabilities of robots. These advancements allowed robots to learn from their environments and adapt their actions based on real-time data. This evolution was a game changer. Robots were no longer confined to repetitive tasks; they began to exhibit decision-making capabilities and could perform more complex operations autonomously.

The rise of humanoid robots in industrial settings is the latest chapter in this evolution. Unlike their predecessors, humanoid robots are designed to work alongside humans, often mimicking human movements and behaviors. They are equipped with advanced sensory systems, enabling them to perceive their surroundings in ways traditional industrial robots could not. This humanoid form factor is significant because it allows for seamless integration into human-centric workspaces without extensive reconfiguration of existing setups.

One of the fascinating aspects of humanoid robots is their potential to bridge the gap between human and machine collaboration. Traditionally, industrial robots have been isolated from human workers due to safety concerns and operational differences. However, humanoid robots are built with advanced safety features and are programmed to understand and respond to human presence, making them ideal collaborators on the factory floor.

Today, robots like Boston Dynamics' Atlas and SoftBank's Pepper are pushing the boundaries of what industrial robots can achieve. Atlas, for example, is not only capable of performing high precision tasks but can also navigate complex environments, identify objects, and even handle unexpected obstacles. This level of sophistication is underpinned by robust AI algorithms that allow these humanoids to process vast amounts of data in real time, making informed decisions instantaneously.

The introduction of AI-powered humanoid robots is not just about task efficiency but also about enhancing the overall work environment. These robots are increasingly being utilized for tasks that require a high degree of dexterity and precision, tasks that have traditionally been challenging to automate. For example, in electronics manufacturing, where handling small and delicate components is crucial, humanoid robots are proving to be invaluable.

Moreover, the evolution of robotics in manufacturing is heavily influenced by advancements in sensor technology and data interpretation. Modern robots are equipped with a plethora of sensors – from cameras and LIDAR systems to tactile sensors – that enable them to gather detailed information about their operating environment. This sensory data is then processed using sophisticated AI algorithms, allowing the robots to react to changes in real-time and perform their tasks with unprecedented efficiency and safety.

The role of software in the evolution of robotics cannot be overstated. Advances in software architectures have enabled the creation of operating systems specifically designed for robotics. These operating systems facilitate real-time data processing, critical for the efficient functioning of humanoid robots. The ability to process and interpret complex data streams in real-time enables these robots to navigate dynamic environments, avoid obstacles, and even predict potential hazards.

As industries continue to adopt humanoid robots, the lessons learned from the evolution of robotics in manufacturing will prove invaluable. One of these lessons is the importance of continuous improvement and adaptation. The success of early industrial robots was built on incremental advancements – small, continuous improvements that collectively led to significant leaps in capabilities. This approach remains relevant today as AI-powered humanoids are constantly being refined and upgraded.

Another critical insight from the journey of robotics in manufacturing is the importance of human-robot collaboration. Early robots were designed to replace human workers for specific tasks. In contrast, modern humanoid robots are designed to work alongside humans, complementing their capabilities and enhancing overall productivity. This collaborative approach is pivotal in addressing some of the key challenges faced by the manufacturing industry, such as labor shortages and the demand for higher productivity.

The evolution of robotics in manufacturing is also a narrative of overcoming challenges and pushing boundaries. Each phase of this journey has been marked by significant technical challenges – from creating reliable mechanical systems to developing sophisticated AI algorithms. The relentless pursuit of solutions to these challenges has driven innovation and accelerated the pace of technological advancement.

The application of humanoid robots in industry is not without its challenges. Issues such as system integration, cost of implementation, and workforce adaptation need to be meticulously addressed. However, the benefits far outweigh the hurdles. Humanoid robots are not only enhancing productivity but are also paving the way for more flexible and adaptive manufacturing processes. This will ultimately lead to more resilient and competitive industries on a global scale.

In conclusion, the evolution of robotics in manufacturing has laid a solid groundwork for the rise of humanoid robots. From simple, task-specific machines to sophisticated AI-powered humanoids, each phase of this journey represents significant strides in technological innovation and human ingenuity. As we witness the integration of humanoid robots in industrial settings, it's clear that we are on the cusp of a new era in manufacturing – one that promises unprecedented levels of efficiency, safety, and collaboration between humans and robots.

Case Study: Early Humanoid Robots

The journey towards integrating humanoid robots in industrial settings began with a significant leap in innovation and vision. Early humanoid robots were not just about mimicking human physicality but also embedding a semblance of human-like intelligence and responsiveness. These pioneering models set the ground for what was to come, showing us both the potential and the limitations of this technology in manufacturing environments.

In the mid-20th century, the first wave of robotic automation primarily involved simple, repetitive tasks. However, the emergence of humanoid robots marked a shift from basic automation to advanced, adaptive functionality. These early humanoid robots, such as WABOT-1 developed by Waseda University in 1973, achieved astounding milestones. WABOT-1 could walk, grip objects, and even communicate in Japanese, albeit in a basic form. This was revolutionary, showcasing the feasibility of robots that could perform tasks in a human-like manner.

Honda's ASIMO, introduced in 2000, represented another crucial development. ASIMO, with its advanced locomotion capabilities, could run, climb stairs, and navigate complex environments, attributes critical for industrial applications. While it was primarily designed as a research tool, ASIMO highlighted the practical potential of humanoid robots, especially in settings requiring high mobility and adaptability. The creation of ASIMO demonstrated how integrating sensors, actuators, and control algorithms could create a robot capable of dynamic and interactive tasks.

Though initially expensive and somewhat limited, these early robots hinted at immense possibilities. Their presence in factories, though more experimental at first, began to inform us about the future landscape of industrial automation. Engineers and scientists gathered key insights on motion control, power management, and human-robot interaction. These early robots served as both functional machines and experimental platforms, pushing the boundary of what could be achieved.

Another significant contender in early humanoid robotics was Kawada Robotics' HRP series. HRP-2, unveiled in 2003, was designed specifically for industrial applications. Unlike ASIMO, HRP-2 had a more utilitarian form factor, modeled closely to fit inside manufacturing environments. It could perform tasks such as welding, assembling, and transporting components, highlighting how humanoid robots could supplement human workers in manipulating objects and operating machinery.

The HRP series taught us valuable lessons in collaborative robotics. For instance, it was evident that humanoid robots could act as intermediaries in environments where automated machinery is too rigid and human intervention too costly. The collaborative nature of these robots opened up new paradigms in industrial automation - robots as assistants rather than mere tools. This marked a significant cultural and operational shift, setting the stage for future human-robot collaboration.

Despite these advancements, early humanoid robots faced substantial challenges. One major issue was energy efficiency. The battery technology at the time limited the operational hours of these robots, restricting their practical use in an industrial context. Additionally, the complexity of mobility and agility algorithms meant that robots often needed substantial downtime for maintenance and recalibration. These technical hurdles necessitated ongoing research and development in both hardware and software domains.

Software also posed a significant challenge. Early humanoid robots were not as sophisticated in learning and interacting. Their decision-making processes were often hard-coded and lacked the adaptability seen in more modern AI-powered systems. Incorporating machine learning and AI would later prove crucial, but these early efforts laid the foundation for understanding the importance of adaptive control systems and real-time data processing in automation.

Examples like the KUKA Heavy Duty robots, although not strictly humanoid, provided insights that would trickle down into humanoid robotics. Industrial robots such as these set benchmarks for speed, precision, and robustness, influencing the design and functionality of early humanoid versions. The lessons drawn from KUKA robots were instrumental in understanding how humanoid robots could be tailored for heavy-duty tasks, particularly in demanding sectors like automobile manufacturing.

Early humanoid robots also contributed to discussions about safety and ergonomics. Deploying these robots required factories to reconsider their layouts to ensure safety for both human workers and the machines themselves. The development of safety protocols and ergonomic designs to accommodate humanoid robots led to safer and more efficient workspaces. These considerations are vital as we continue to integrate more sophisticated humanoid robots into various industrial settings.

It is important to acknowledge the visionaries behind these machines. Early pioneers, through sheer determination and creativity, sculpted the pathway for modern humanoid robotics. Their interdisciplinary approach combined insights from fields as diverse as mechanical engineering, computer science, cognitive science, and even philosophy. They envisioned a future where robots could seamlessly work alongside humans, pushing the boundaries of productivity and innovation.

As we look back, the evolution of these early humanoid robots serves as an enlightening case study. It illustrates the progression from rudimentary automation to sophisticated, AI-powered systems capable of transforming industries. Every iteration brought us closer to the seamless integration of humanoid robots in the industrial workforce, enhancing both efficiency and safety. This historical context is not just about technological evolution; it's about understanding the steps and missteps that pave the way for future advancements.

In conclusion, early humanoid robots played a critical role in the ongoing saga of industrial automation. While their functionalities were limited compared to today's standards, their impact was profound. They bridged the gap between imagination and reality, turning science fiction into practical engineering. As we build on their legacy, the lessons learned from these pioneering machines continue to guide us, pushing the envelope of what's possible in factory automation.

Chapter 2:
AI and Machine Learning in Robotics

In the ever-evolving landscape of industrial automation, AI and machine learning have emerged as game-changers, fueling unprecedented advancements in robotics. These cutting-edge technologies allow humanoid robots to not only perform tasks with remarkable precision but also to learn and adapt in real time. Gone are the days when robots were confined to static, repetitive motions; today, they can analyze vast amounts of data, make intelligent decisions, and even predict potential failures before they occur. This transformative capability augments the efficiency of factory operations, bringing a new level of agility and resilience to the production floor. The integration of AI into robotics accelerates innovation and redefines our expectations of what machines can accomplish. As manufacturers continue to harness these powerful tools, the boundary between human ingenuity and robotic efficiency blurs, setting the stage for a new era of collaborative and intelligent industrial environments.

Core Technologies Driving Humanoids

Artificial intelligence (AI) and machine learning are the beating heart of modern humanoid robots. While traditional robots have long been a staple in industrial settings, their capabilities pale in comparison to their AI-powered counterparts. The ability of humanoid robots to adapt, learn, and make decisions in real-time significantly enhances their utility in diverse manufacturing environments.

A key element in the advancement of humanoid robots is deep learning. This subset of AI employs neural networks with several layers to analyze complex patterns in data. By simulating the human brain's approach to problem-solving, deep learning enables robots to perform tasks that were previously thought to be exclusively within the human domain. For example, deep learning allows robots to inspect products for defects with a level of precision and nuance that manual inspections often miss.

Another cornerstone technology is natural language processing (NLP). This allows humanoid robots to understand and generate human language. In an industrial setting, NLP facilitates seamless communication between humans and robots, simplifying tasks that require verbal instructions or vocal feedback. For instance, factory workers can easily communicate adjustments or instructions without needing specialized programming skills, leading to more intuitive human-robot interactions.

Computer vision is equally transformative. Equipped with high-resolution cameras and sophisticated image recognition algorithms, humanoid robots can identify objects, navigate environments, and even detect emotional cues in human faces. This capability is critical for tasks that require a high level of precision and safety, such as assembling intricate components or working alongside human colleagues in confined spaces.

Reinforcement learning, another vital AI technique, teaches robots to improve their performance through trial and error. This approach mimetitizes the way humans learn from experience, allowing robots to autonomously enhance their skillset over time. For instance, a robot responsible for welding might start with basic proficiency but gradually fine-tune its performance to achieve near-perfect welds, thus improving the quality and consistency of the final product.

Advanced sensor technologies play an equally crucial role. These sensors provide real-time feedback on various parameters such as proximity, temperature, and pressure. When integrated with machine learning algorithms, the sensor data allows humanoid robots to make immediate adjustments, enhancing their accuracy and efficiency. For example, pressure sensors in a robotic gripper can ensure delicate items are handled without damage, while proximity sensors help the robot avoid collisions in dynamic environments.

Another monumental leap in humanoid robotics is the development of decision-making algorithms. These algorithms enable robots to assess various courses of action and choose the most efficient path based on predefined goals and current conditions. While traditional robots follow rigidly pre-programmed paths, AI allows for a more flexible and situationally-aware approach. This adaptability is particularly beneficial in manufacturing contexts where unexpected changes and challenges are the norm rather than the exception.

Edge computing has also revolutionized the way data is processed in humanoid robots. By analyzing data at the point of collection, rather than relying solely on the cloud, edge computing reduces latency and improves response times. This capability is invaluable in scenarios that demand real-time decision-making, such as quality control inspections or emergency shutdowns. Faster data processing ensures that robots can react almost instantaneously, thus reducing errors and enhancing productivity.

Robotics operating systems (ROS) provide the software infrastructure necessary for the development and integration of complex robotic systems. These systems facilitate communication between the robot's various components and simplify the overall design and programming process. ROS enables modularity, allowing for easy updates and upgrades to both hardware and software. This flexibility ensures that humanoid robots remain at the cutting edge of technological advancements and can be swiftly adapted to new tasks and environments.

In addition to these technologies, collaboration mechanisms are integral to the operation of humanoid robots in industrial settings. Collaborative robots, often referred to as cobots, are designed to work alongside human operators safely and efficiently. Sophisticated motion control algorithms and safety mechanisms ensure that cobots can operate in close proximity to humans without posing any risks. This synergy between humans and robots not only improves productivity but also enhances job satisfaction by reducing the physical strain on human workers.

Another innovation critical to humanoid robots is the incorporation of exoskeletons. These wearable devices support and amplify human movement, allowing for the execution of tasks that require significant strength or endurance. In an industrial setting, exoskeletons can be paired with humanoid robots to create powerful human-robot hybrid teams capable of extraordinary feats of productivity.

Lastly, simulation and digital twin technologies provide a virtual representation of physical robots and their working environments. These digital models allow for extensive testing and training in a risk-free virtual environment before implementation in the real world. For example, engineers can use digital twins to simulate an entire manufacturing line, identifying bottlenecks and optimizing processes before actual deployment. This preemptive approach significantly reduces downtime and implementation risks.

Collectively, these core technologies constitute the driving force behind the new generation of humanoid robots. They don't just perform tasks; they learn, adapt, and interact in ways that are fundamentally reshaping the future of industrial automation. The rapid advancements in AI and machine learning promise a future where manufacturing processes are not only more efficient but also more intelligent and adaptive. As these technologies continue to evolve, the potential for humanoid robots to revolutionize industrial production grows exponentially, making them indispensable assets in the modern factory landscape.

Differences Between Traditional and AI-Powered Robots

When you think about robots in industrial settings, the first image that might come to mind is one of rigid, pre-programmed machines performing repetitive tasks with unerring precision. These are traditional robots, and they've been an integral part of manufacturing for decades. However, the landscape is rapidly changing with the advent of AI-powered robots, which bring a whole new level of capability and flexibility to the factory floor.

Traditional robots operate based on precise, hard-coded instructions. They excel at repetitive tasks in controlled environments, where variability is minimized. Their job, whether it's welding car parts or assembling electronics, is meticulously pre-defined by human programmers. This inflexibility means that even slight changes in the environment or task may require substantial reprogramming. This rigidity is beneficial for certain applications, yet it doesn't lend itself well to dynamic and complex scenarios.

AI-powered robots, on the other hand, are a different breed altogether. Powered by machine learning algorithms, these robots can learn from their environment and experiences, making them far more adaptable. They can process a range of data inputs—from visual to auditory—and adjust their actions accordingly. It's not just about doing a task over and over again; it's about doing it better each time.

The impressive part is that these robots get smarter and more efficient with every iteration.

Imagine a traditional assembly line, where a robot's malfunction could halt production. In such cases, a technician would need to diagnose the issue, possibly reprogram the robot, and restart the production line—a time-consuming process. Now, consider an AI-powered robot in the same scenario. It would use predictive analytics to anticipate the malfunction and either correct its own operations or alert a human operator before the failure occurs. This kind of proactive problem-solving significantly reduces downtime and increases productivity.

Another stark difference lies in their interaction capabilities. Traditional robots are confined by their programming, which means they don't interact in any meaningful way with human workers. Safety protocols usually restrict human-robot interactions to avoid accidents. Conversely, AI-powered robots come with advanced sensory systems and autonomous decision-making abilities. They can safely work alongside humans, understanding and reacting to human gestures, language, and even emotions to some extent. This collaborative potential opens avenues for more integrated and efficient workflows.

The flexibility of AI-powered robots is another game changer. Traditional robots are generally designed and set up for specific tasks, and transitioning them to new roles can be cumbersome. By contrast, AI-powered robots can be re-trained for different tasks without extensive reprogramming. This adaptability proves invaluable in industries where production requirements are constantly evolving, allowing businesses to remain agile and responsive to market demands.

Furthermore, the types of tasks that AI-powered robots can handle are expanding rapidly. Traditional robots have been limited to jobs that are too dangerous, dirty, or dull for humans. While these criteria still apply, AI-powered robots are increasingly capable of performing intricate tasks that require a level of subtlety and decision-making previously thought to be the domain of human workers. From quality control to complex assembly tasks, their utility spans a broader spectrum.

What's truly revolutionary about AI-powered robots is their ability to innovate autonomously. Through machine learning, they can optimize their processes, discern patterns invisible to the human eye, and even suggest improvements to the tasks at hand. Think of an AI-powered robot that, after months of assembling a product, identifies a more efficient assembly technique and implements it on its own. This level of innovation is unheard of with traditional robots.

The data-driven nature of AI-powered robots also brings a layer of intelligence to their operation. They constantly collect data—from their performance metrics to environmental conditions—and analyze it to improve their efficiency. This real-time feedback loop makes them incredibly responsive and capable of continuous self-improvement. Traditional robots, lacking such sophisticated data integration, remain static, performing the same tasks in the same way until a human decides otherwise.

Yet, it's not just about the robots themselves; it's also about the ecosystems they create. Traditional robots require highly structured environments to function optimally. Everything around them must be meticulously organized to fit their limited operational parameters. AI-powered robots bring in a different paradigm. They can adapt to unstructured environments and work with varied components and processes without requiring extensive changes to the production line. This adaptability means factories can be more versatile and less rigid in their design and operations.

Safety is another critical area where AI-powered robots outshine their traditional counterparts. Built-in AI systems enable these robots to detect and respond to safety hazards in real time. They can anticipate potential risks and take corrective actions, such as shutting down operations or alerting human supervisors. These advanced safety features considerably reduce the likelihood of accidents, fostering a safer working environment.

Moreover, AI-powered robots are better at handling exceptions and anomalies. Traditional robots falter when faced with unexpected conditions; they typically cease operations until a human intervenes. AI robots, however, can recognize anomalies and adjust their actions accordingly, minimizing disruptions. For instance, if an AI robot detects a defect in a material, it can flag the issue and reroute the item for inspection without halting the entire production line.

Another aspect worth mentioning is maintenance. Traditional robots follow a time-based maintenance schedule, meaning they receive periodic maintenance regardless of their actual condition. AI-powered robots, however, are capable of predictive maintenance. They monitor their own performance in real-time and identify when maintenance is actually needed. This targeted approach not only reduces downtime but also extends the life of the robots by preventing unnecessary wear and tear.

In terms of economic impact, the shift from traditional to AI-powered robots is profound. Initial costs for implementing AI robots can be higher, mainly due to the complexity of their systems and the advanced technologies involved. However, the long-term benefits—such as increased productivity, reduced downtime, and lower maintenance costs—often outweigh the initial investment. Companies find that the return on investment for AI-powered robots is considerably higher over time.

Lastly, let's consider future-proofing. Traditional robots, with their limited adaptability, can quickly become obsolete as manufacturing techniques and products evolve. AI-powered robots, designed for learning and adaptation, offer longevity. They can evolve alongside technological advancements, making them a more sustainable investment for the future.

In summary, the transition from traditional to AI-powered robots represents a leap forward in industrial automation. Traditional robots, while reliable and efficient in their own right, lack the flexibility, adaptability, and intelligence that AI-powered robots bring to the table. As industries strive for greater efficiency, precision, and innovation, AI-powered robots are uniquely positioned to meet these challenges, paving the way for a new era in manufacturing.

Chapter 3:
Automotive Industry: A Pioneer in Robotics

In the realm of industrial automation, the automotive industry has undoubtedly stood as a frontrunner, embracing the integration of humanoid robots to revolutionize production lines. This industry's early adoption of robotics, driven by the unyielding quest for efficiency and precision, has set benchmarks across manufacturing sectors. Automotive plants were among the first to identify and harness the potential of humanoid robots, incorporating them into complex assembly processes that once relied solely on human dexterity and decision-making. The transformative impact on production speed and quality has been monumental, leading to significant reductions in error rates and production costs. As a result, these advancements have not only redefined industry standards but have also inspired other sectors to explore the boundless possibilities of AI-powered robotics.

First Humanoid Integrations in Automotive Plants

The automotive industry, known for its early adoption and continuous innovation in robotics, took a monumental leap forward with the integration of humanoid robots in its manufacturing plants. Traditionally, the sector relied on industrial robots for tasks that required high precision and repetitive actions, such as welding and painting. However, the introduction of humanoid robots marked a significant shift, as these machines brought new levels of dexterity, adaptability, and collaborative abilities that were previously unattainable.

Early adopters of humanoid robots in the automotive industry viewed these machines as not just replacements for human labor but as co-workers capable of seamlessly integrating into existing workflows. These humanoids were designed to mimic human actions closely, making them ideal for complex, non-repetitive tasks that require a certain level of human-like judgment and adaptability. For instance, humanoid robots could handle tasks such as aligning parts during assembly, which previously needed the nuanced touch of a skilled human worker.

The bridging of this gap between human capabilities and traditional robotic efficiency highlighted the transformative potential of humanoid robots.

The first integration of humanoid robots in automotive plants faced numerous challenges, from technological limitations to workforce acceptance. Early models had to continuously undergo rigorous testing and refinement to meet the high standards required in automotive manufacturing. Unlike traditional robots, which could be programmed for specific, repetitive tasks, humanoid robots had to learn and adapt to varying situations, which necessitated advanced AI algorithms and real-time data processing capabilities. This ongoing evolution was essential to ensure the robots could function effectively alongside human workers without causing disruptions.

One of the pivotal moments in this journey was when Japan's leading automotive manufacturer, Toyota, introduced its first-generation humanoid robots in their assembly lines. These robots were equipped with sophisticated sensors, flexible limbs, and intuitive AI, allowing them to perform tasks such as screwing and component fitting with minimal human intervention. The success of this implementation demonstrated the feasibility and benefits of humanoid robots, encouraging other industry players to explore similar integrations.

Beyond addressing labor shortages and reducing operational costs, humanoid robots brought about a substantial improvement in production speed and product quality. Their ability to work tirelessly and with consistent precision minimized the margin for error, leading to higher quality outputs. Moreover, these robots could operate 24/7 without the need for breaks, thus significantly increasing overall plant productivity. The automotive industry, driven by fierce competition and demand for high-quality products, found in humanoid robots a viable solution to meet these challenges head-on.

Another striking advantage of humanoid integration was the enhanced safety it provided in the workplace. These robots were deployed in environments and tasks that posed potential hazards to human workers, such as handling toxic substances or operating in extreme conditions. Their introduction not only reduced the incidence of workplace accidents but also relieved human workers from monotonous and physically demanding tasks, hence improving overall job satisfaction and occupational well-being.

Looking at the broader picture, the adoption of humanoid robots in automotive plants fostered a culture of innovation and continuous improvement. Companies started investing heavily in research and development to further push the boundaries of what these robots could achieve. Collaborative robots, or "cobots", became a common term, reflecting a shift towards workplaces where humans and robots could work together in harmony. This focus on collaboration extended to the design of workspaces, which were increasingly optimized to accommodate both human and robot workers effectively.

The journey of integrating humanoid robots into automotive manufacturing has not been without its challenges. Educational programs and training were pivotal in ensuring that the existing workforce could smoothly transition to working alongside these new robotic colleagues. Workshops and seminars were held to familiarize employees with the operational aspects of humanoid robots, addressing concerns and uncertainties. This proactive approach was crucial in mitigating resistance to change and fostering a culture of acceptance and cooperation.

Institutional barriers also had to be overcome. Policies and regulations regarding the deployment of humanoid robots in industrial settings were in their infancy and needed comprehensive revisions. Companies had to navigate through a complex landscape of compliance requirements, ensuring both safety and efficacy in operations involving humanoid robots. To tackle these hurdles, many companies started working closely with regulatory bodies, contributing valuable insights to the development of more robust and forward-thinking guidelines.

Humanoids' impact wasn't just confined within the factory walls; it rippled outwards, influencing supply chains and logistics. These robots, equipped with advanced AI and machine learning capabilities, could predict and adapt to changes in supply and demand more swiftly than traditional systems. For instance, they could optimize inventory management, ensuring that parts and materials were available just in time for production, thereby reducing downtime and enhancing overall efficiency.

As the first wave of humanoid robots proved their mettle in the automotive industry, their role evolved from being mere tools to becoming integral components of the manufacturing ecosystem. Continuous improvements in AI and sensor technologies kept enhancing their capabilities, allowing them to perform even more sophisticated tasks. This ongoing evolution was not only about adaptation but also about anticipating future needs and staying ahead of the curve in a rapidly changing industrial landscape.

The successful integration of humanoid robots in automotive plants set the stage for broader applications across various industries. The automotive sector became a case study in how humanoid robots could revolutionize manufacturing, offering invaluable insights and guiding principles for other sectors aiming to leverage this transformative technology. The lessons learned from these pioneering efforts provided a blueprint for overcoming initial obstacles and maximizing the benefits of humanoid robots in industrial settings.

Ultimately, the first humanoid integrations in automotive plants signaled a new era in manufacturing, where the symbiosis between human ingenuity and robotic precision could drive unprecedented levels of efficiency and innovation. This transformation underscored the automotive industry's role as a pioneer in robotics, pushing the boundaries of what's possible and inspiring other sectors to follow suit. The exploration of this synergy continues, shaping a future where humanoid robots are not just tools but partners in the relentless pursuit of excellence.

Impact on Production Speed and Quality

The integration of AI-powered humanoid robots in the automotive industry has significantly impacted production speed and quality. One of the most notable improvements is the reduction in cycle times. Humanoid robots bring a level of precision and efficiency that traditional robotic systems and human workers simply can't match. Advanced algorithms and real-time processing capabilities allow these robots to perform complex tasks faster and with greater accuracy.

For instance, in welding applications, the precision of humanoid robots ensures that welds are consistently exact, reducing the likelihood of defects and rework. This leads to a higher-quality end product and less wastage of materials. Moreover, the speed at which these robots can work drastically shortens production cycles. Where a human worker might take several minutes to complete a task, a humanoid robot can complete it in a fraction of the time without compromising on quality.

An automotive assembly line is a perfect example of where humanoid robots shine. Tasks such as installing seats, connecting electrical wiring, and even painting have been enhanced by the introduction of these robots. Their AI-driven systems can adapt to different models without requiring extensive reprogramming. This flexibly means that mixed-model production—where different models of cars are produced on the same line—can proceed without the delays traditionally associated with changeovers.

Humanoid robots are also contributing to improved quality assurance processes. Equipped with advanced sensors and machine learning algorithms, they can identify defects at earlier stages, drastically reducing the number of faulty end products. This kind of predictive quality control allows for immediate adjustments in the manufacturing process, preventing issues from escalating. These improvements in quality assurance not only enhance the final product but also build consumer trust and satisfaction, which are crucial in the competitive automotive market.

In terms of material handling, humanoid robots bring unmatched dexterity and efficiency. With their ability to perform repetitive and physically demanding tasks without fatigue, they ensure a continuous, smooth workflow. Their precision reduces the risk of material damage, further contributing to overall quality. This is particularly important in the handling of delicate components such as electronic modules and glass installations, where even minor mishandling can result in significant quality issues.

In addition to precision, the enhanced speed of humanoid robots leads to substantial cost savings. Shorter production times mean that more units can be produced within the same period, increasing output without the need for additional shifts. This boosts the plant's overall productivity and profitability. For example, companies have reported up to a 20% increase in production capacity after integrating humanoid robots into their assembly lines. These efficiencies translate into reduced lead times, allowing automakers to meet market demands more swiftly.

Furthermore, the deployment of humanoid robots also has significant implications for the consistency of the production process. Robots don't tire, need breaks, or suffer from the variability inherent in human labor. Their unwavering reliability ensures that every task is performed to the same high standard, every time. This consistency is particularly beneficial in highly regulated industries like automotive manufacturing, where stringent compliance with safety and quality standards is non-negotiable.

The impact on production speed is also evident in the realm of maintenance and downtime. Traditional production lines often suffer from unforeseen downtimes due to equipment failure or human error. However, humanoid robots equipped with predictive maintenance technologies can foresee potential issues before they become critical, enabling proactive maintenance. This reduces downtime significantly, ensuring that production lines run smoothly and efficiently with fewer interruptions.

Moreover, the use of AI-powered systems in humanoid robots allows for continuous learning and improvement. These robots can analyze and adapt to new data from their environment in real-time. In the context of production speed and quality, this means they are continually optimizing their performance, finding efficiencies, and improving the processes they are involved in. Such dynamic capabilities are a step beyond the rigid programming of traditional robots and human operations, offering an adaptive advantage that directly impacts productivity.

The positive ripple effect of integrating humanoid robots extends beyond the immediate production floor. Faster production cycles and higher quality outputs can reduce the time to market for new automotive models, giving companies a competitive edge. As product lifecycles grow shorter due to rapid technological advancements and shifting consumer preferences, the ability to accelerate production without sacrificing quality is an invaluable asset for automakers.

Finally, it's crucial to consider the human element in this transformation. While robots take over the repetitive and hazardous tasks, human workers can be reassigned to more strategic roles that require critical thinking and creativity. This not only improves job satisfaction but also augments the overall quality of the workforce. Companies investing in upskilling their employees alongside deploying humanoid robots are finding that the synergy between human and machine creates a more dynamic, innovative, and engaged production environment.

In summary, the impact of AI-powered humanoid robots on production speed and quality within the automotive industry is profound. They enhance precision, reduce cycle times, improve consistency, and ensure a higher standard of quality. These improvements lead to cost savings, reduced lead times, and a stronger competitive position in the market. As technology continues to evolve, the role of humanoid robots in driving production efficiencies and maintaining high-quality standards will only become more pivotal.

Chapter 4:
Key Components of Humanoid Robots

Understanding the key components of humanoid robots is crucial for comprehending their transformative impact on industrial automation. The core elements include advanced AI algorithms and decision systems that empower these robots to perform complex tasks with precision and adaptability. Sensor technologies are pivotal as they enable robots to perceive their surroundings through an array of sensors, interpreting data in real time to navigate and interact within dynamic environments. Power systems and actuators provide the necessary movement capabilities, mimicking human dexterity and strength. Additionally, robust communication interfaces ensure seamless interaction between humans and robots, fostering collaborative efforts on the factory floor. Together, these components work in harmony to create sophisticated, efficient, and responsive humanoid robots that are revolutionizing production processes and enhancing overall operational efficiency.

AI Algorithms and Decision Systems

AI algorithms and decision systems form the brain behind humanoid robots in industrial settings. These algorithms are intricate frameworks that enable robots to make decisions in real-time, adapt to new data inputs, and optimize processes for efficiency and productivity. The sophistication of these systems allows robots not just to perform tasks but to understand their environment and make judgement calls similar to human workers.

At the core of these decision systems are machine learning models. These models enable humanoid robots to learn from past experiences, adjust their actions based on outcomes, and improve their performance over time. Supervised learning, where the robot is trained on a labeled dataset, and unsupervised learning, which allows the robot to identify patterns and relationships from unstructured data, are both pivotal in this context. Reinforcement learning, a subtype of machine learning, is particularly crucial as it allows robots to learn optimal strategies through trial and error, followed by rewarding desirable behaviors.

For instance, consider a humanoid robot tasked with assembling vehicle components on an automotive production line. Using reinforcement learning, the robot iteratively improves its assembly techniques by receiving feedback—be it through error measurement or productivity scores. This constant loop of action and feedback significantly boosts its efficiency, mirroring the incremental learning curve that human workers experience over time.

Complementing machine learning models are deep learning algorithms, which drive the robot's ability to interpret complex data inputs such as visual, auditory, and sensor data. Deep neural networks with multiple layers allow for the extraction of features from raw sensor data. This capability is essential when robots need to navigate around obstacles, identify objects, or perform intricate tasks like welding or painting with high precision. The development of convolutional neural networks (CNNs) has been particularly transformative in image recognition, enabling robots to visually inspect products for quality assurance.

Natural language processing (NLP) algorithms are another integral component. These algorithms empower humanoid robots to understand and respond to verbal commands from human co-workers, making human-robot collaboration more seamless. NLP not only helps in command interpretation but also in documenting and logging conversations for future analysis. Chatbot interfaces used in customer service are a practical application of NLP, and similar technologies are now being embedded in industrial humanoid robots to provide real-time support and troubleshooting advice.

Moreover, the integration of AI algorithms with decision-making frameworks like Markov decision processes (MDPs) and Bayesian networks enhances a robot's ability to predict outcomes and make probabilistic decisions. MDPs are particularly effective in scenarios where the robot must make a series of decisions that can lead to different probabilistic outcomes. Bayesian networks, on the other hand, provide a way to model the uncertainty in various scenarios, allowing the robot to make informed decisions even with incomplete data.

In real-time applications, the speed and accuracy of decision-making are paramount. This is where digital twins come into play. A digital twin is a virtual model of a physical robot and its operating environment. By simulating real-world conditions, digital twins allow for the testing and refinement of AI algorithms in a risk-free virtual setting. Once optimized, these algorithms can be deployed to physical robots, ensuring high-performance standards from the get-go. This approach minimizes downtime and maximizes efficiency by addressing potential issues before they manifest in the real world.

Additionally, ensemble learning techniques—where multiple AI models are used in conjunction—provide robust decision-making capabilities. These techniques aggregate the predictions of multiple models to improve overall accuracy and reliability. For example, if one model is optimized for speed but another for accuracy, their combined predictions offer a balanced outcome, making the robot's decisions more reliable and efficient.

Another fascinating development is the use of swarm intelligence in AI decision systems. Inspired by the collective behavior of social insects like bees or ants, swarm intelligence enables a group of robots to work together toward a common goal. Through simple rules and local interactions, the collective behavior of these robots can solve complex problems that would be difficult for a single robot to tackle. This technique is particularly useful in supply chain management and logistics, where multiple robots can coordinate to optimize warehouse operations.

It's crucial to note that the effectiveness of these AI algorithms and decision systems hinges on robust data collection and preprocessing. High-quality data is the lifeblood that fuels these algorithms, ensuring their decisions are reliable and trustworthy. Advanced sensor technologies, which will be discussed in the subsequent sections, play a pivotal role in gathering this data. Proper data preprocessing—such as normalization, handling missing values, and feature extraction—ensures the data fed into AI models is accurate and meaningful.

The continual improvement of AI algorithms and decision systems is also supported by real-time analytics and feedback loops. Integration with industrial Internet of Things (IIoT) platforms allows humanoid robots to receive continuous streams of data from various sources, facilitating real-time decision-making. These IIoT platforms also enable predictive analytics, where historical data is used to predict future events, further enhancing the decision-making process.

Furthermore, the ethical considerations surrounding AI decision systems cannot be overlooked. Ensuring transparency in decision-making processes is paramount to gain the trust of human co-workers. Explainable AI (XAI) techniques are being developed to make the decision-making processes of humanoid robots more transparent and understandable to humans. By providing clarity on how a particular decision was reached, XAI fosters a collaborative working environment where humans and robots can coexist harmoniously.

Lastly, the integration of AI algorithms with cloud computing and edge AI, as discussed in later chapters, allows for scalable and efficient processing power. Cloud-based AI platforms provide vast computational resources, enabling the training of complex models on large datasets, while edge AI facilitates real-time decision-making by processing data locally on the robot. This hybrid approach ensures that humanoid robots can handle computationally intensive tasks without latency issues while maintaining high responsiveness.

In conclusion, AI algorithms and decision systems are the lynchpin of modern humanoid robots. They transform these machines from mere executors of predefined tasks to dynamic entities capable of learning, adapting, and making intelligent decisions. As industrial automation continues to evolve, the sophistication of these AI systems will only deepen, driving even greater efficiencies and innovation in manufacturing environments.

Sensor Technologies and Data Interpretation

Humanoid robots in industrial settings are increasingly reliant on advanced sensor technologies to navigate dynamic environments and perform tasks with precision. These sensors are the eyes, ears, and sometimes even the skin of the robots, enabling them to collect a plethora of data in real-time. This data is indispensable for the robot's decision-making algorithms and helps in interpreting complex scenarios that are inherent in manufacturing processes.

One of the cornerstone sensors employed in humanoid robots is LiDAR (Light Detection and Ranging). Using laser pulses, LiDAR helps robots create detailed 3D maps of their surroundings. This spatial awareness is crucial for tasks like moving through crowded factory floors or picking items from a conveyor belt with pinpoint accuracy. Additionally, cameras, both RGB and infrared, provide visual data that complements the information gathered by LiDAR. These cameras are not just for object detection but also for recognizing colors, textures, and even reading barcodes or labels on components.

Proximity sensors and ultrasonic sensors serve another critical role, particularly in ensuring the safety of human workers in a mixed environment. These sensors allow humanoid robots to detect obstacles that may not be visible through cameras, such as transparent or reflective surfaces. When an obstacle is detected, the robot can either slow down or reroute itself to avoid collisions, thus preventing accidents and maintaining a seamless workflow.

Amongst the most sophisticated types of sensors are force and tactile sensors. Embedded in the robot's joints and fingertips, these sensors enable the robot to gauge how much force to apply when handling different materials, from delicate electronic components to heavy automotive parts. Essentially, they give the robot a sense of "touch," making it possible to perform complex assembly tasks that require a delicate balance of strength and finesse.

Interpreting the data collected by these sensors relies heavily on AI algorithms and machine learning models. The wealth of sensor data can be overwhelming, but advanced data interpretation systems categorize, analyze, and draw actionable insights in real time. This, in turn, allows humanoid robots to adapt to changing conditions on the factory floor. For instance, if a robot equipped with temperature sensors detects that a machine is overheating, it can alert human supervisors or even take pre-programmed actions to mitigate the risk.

The integration of advanced sensor technologies extends even further with the utilization of environmental sensors. These sensors monitor factors such as humidity, air quality, and temperature within the factory. This contextual awareness is particularly beneficial in industries like pharmaceuticals or food processing, where maintaining specific environmental conditions is essential for product quality and safety.

Robotic systems also leverage proprioceptive sensors to understand their own state. Gyroscopes and accelerometers fall into this category, offering insights into the robot's position, orientation, and movement. This feedback loop is vital for tasks requiring high degrees of precision and balance, such as stacking items or transferring objects from one location to another. By knowing its own position, the robot can adjust its actions in real time to achieve optimal results.

Furthermore, the interpretive layer of sensor data involves the deployment of neural networks that have been trained to recognize patterns and anomalies. This form of data interpretation goes beyond mere detection; it offers predictive analytics that can foresee potential issues before they become critical. For instance, vibration sensors on manufacturing equipment can predict mechanical failures, allowing for preemptive maintenance that minimizes downtime.

Edge computing plays a notable role in the data interpretation pipeline. By processing data at the point of collection, rather than sending it all to a central cloud server, edge computing ensures lower latency and more immediate responses. This is particularly important in high-speed manufacturing environments where delays, even in milliseconds, can be costly. Edge AI enables robots to make split-second decisions based on real-time data analytic outcomes.

A critical component in the interpretation of sensor data is the human-machine interface (HMI). HMIs provide operators with real-time dashboards and analytical tools that visualize sensor data. These interfaces not only aid in monitoring but also in intervening when necessary. For instance, if an HMI system shows that a robot's battery is running low, a human operator can plan for a recharge or replacement, avoiding any interruptions in the production line.

Networked sensors and the Internet of Things (IoT) further amplify the capabilities of humanoid robots in industrial settings. IoT devices can communicate with each other, creating a cohesive network where both robots and machines share data. This interconnectedness allows for more efficient coordination and automation. Robots can, for instance, communicate with automatic guided vehicles (AGVs) to synchronize material handling tasks, thereby increasing overall productivity.

Interestingly, humanoid robots are beginning to incorporate biometric sensors that can monitor the health and well-being of human workers. These sensors measure parameters such as heart rate and skin conductivity, offering invaluable data that can help in making the workplace safer and more ergonomic. When a worker shows signs of fatigue, the system can recommend breaks or adjustments to reduce the risk of injury.

Ultimately, the rich tapestry of sensor technologies and advanced data interpretation methods form the backbone of humanoid robots' capabilities in modern factories. These systems not only enhance efficiency but also bring a level of sophistication and adaptability that was previously unimaginable. As sensor technologies continue to evolve, the potential applications for humanoid robots in industrial settings will only expand, driving forward the fourth industrial revolution and ushering in a new era of manufacturing excellence.

Chapter 5:
Software Architectures in Humanoid Robots

In the dynamic field of humanoid robotics, the software architecture is the backbone that orchestrates the multifaceted functionalities, ensuring seamless operations and adaptability in complex industrial environments. These architectures transcend conventional software design by integrating specialized operating systems tailored for robotic needs, real-time data processing capabilities, and sophisticated AI algorithms. This advanced software infrastructure enables humanoid robots to process a deluge of sensory data instantaneously, make autonomous decisions, and learn from their interactions. The robustness and flexibility of these software systems are pivotal in harmonizing various components—from sensors to actuators—thus optimizing performance and reliability. Ultimately, the innovation in software architectures propels humanoid robots into becoming indispensable assets in modern manufacturing, driving efficiency and paving the way for a more streamlined and intelligent future in industrial automation.

Operating Systems Tailored for Robotics

As the robotics industry advances, the need for specialized operating systems (OS) tailored to the unique demands of humanoid robots has become increasingly apparent. Traditional operating systems like Windows, macOS, or even general-purpose Linux distributions are insufficient for the complexities involved in robotics. Unlike conventional systems, robotics OS must manage real-time data processing, sensor integration, and advanced AI-driven algorithms effectively.

At the heart of robotics-specific OS is the capacity for real-time operation. Time delays and latencies, even in milliseconds, could be detrimental to the performance of a humanoid robot. This necessitates a system that can prioritize tasks and manage resources with unparalleled precision. Real-time operating systems (RTOS) are designed to meet these demands, offering deterministic performance that ensures actions occur within predictable timeframes.

Furthermore, the integration of diverse sensors, actuators, and computing resources is another challenge tackled by robotics operating systems. These components must communicate seamlessly, allowing the robot to interpret data from its environment accurately and execute corresponding actions. This often involves the implementation of middleware, such as the Robot Operating System (ROS), which provides a standard framework for communication between various hardware and software components.

One of the standout features of ROS, and similar middleware, is its modularity. Developers can write and plug in new software modules without overhauling the entire system. This modular architecture simplifies upgrades and customization, making it easier to adapt robots for various industrial tasks. The ability to integrate new functionalities without significant downtime is crucial for maintaining the constant operation demanded in manufacturing environments.

Moreover, ROS includes powerful tools for simulation, visualization, and debugging. Before a robot interacts with the physical world, developers can test and refine its behaviors in a virtual environment. This capability reduces the risk of malfunctions, ensuring smoother deployments and more reliable performance in real-world settings.

Another essential aspect is the management of AI and machine learning algorithms. Operating systems tailored for robotics need to support intensive computational tasks while balancing power consumption and operational efficiency. AI-driven robots require high-performance computing, but they cannot afford the luxury of massive, power-hungry data centers. Operating systems must therefore be optimized for energy efficiency while maintaining high levels of computational prowess.

Security is yet another critical consideration. As robots become more integrated into industrial environments, they become potential targets for cyber-attacks. Robust security protocols are indispensable to protect sensitive data and ensure the integrity of robotic operations. This includes encryption, regular software updates, and network security measures that guard against unauthorized access and tampering.

In terms of adaptability, these systems also need to support a range of hardware configurations. Unlike consumer electronics, which can stick to a standardized set of hardware, robots may involve custom-built components tailored for specific applications. The OS must therefore be hardware-agnostic, capable of configuring itself according to the underlying hardware without manual intervention.

In industrial settings, the capability to scale operations efficiently is another benefit of specialized operating systems. As factories grow and new robotic platforms are introduced, the operating system must facilitate easy integration and scaling. This adaptability ensures that productivity scales with technological investments, maximizing returns on investment.

Moreover, robotics operating systems often include features for predictive maintenance and diagnostics. These systems can monitor the health of various robot components, predicting failures before they occur. This proactive approach minimizes downtime, enhances operational efficiency, and extends the lifespan of the robots, thus offering long-term economic benefits.

To put it all together, specialized operating systems tailored for robotics are a cornerstone of effective and efficient humanoid robots in industrial environments. They address the unique challenges posed by real-time processing, sensor and actuator integration, AI algorithm management, and security. By providing a robust, flexible, and powerful foundation, these operating systems enable robots to perform complex tasks with reliability and precision, reshaping the future of manufacturing.

As we move forward, continuous advancements in operating systems for robotics will drive further innovations in the field. Enhanced real-time capabilities, more secure and adaptive frameworks, and refined development tools will empower the next generation of humanoid robots. These systems will not only enhance industrial efficiency but also open up new possibilities for automation and human-robot collaboration.

Real-Time Data Processing

One of the most critical aspects of humanoid robots in industrial settings is real-time data processing. This capability allows robots to make instantaneous decisions based on a continuous influx of data, ensuring that operations are not just automated but optimized. In a fast-paced manufacturing environment, where milliseconds can make a significant difference, the capacity for real-time data processing is nothing short of revolutionary.

Real-time data processing gives humanoid robots the ability to analyze and act on information almost instantaneously. This aspect is crucial in various scenarios, from identifying defects in real-time to adjusting processes based on real-time performance metrics. The ability to process data quickly allows these robots to respond to dynamic conditions, ensuring that production lines remain efficient and free of costly interruptions.

This process involves collecting data from multiple sensor inputs, running that data through complex algorithms, and then executing appropriate actions. Sensors on a humanoid robot may include visual cameras, thermal cameras, LIDAR, and more. The data from these sensors is constantly streamed to the robot's processing units, where sophisticated machine learning algorithms analyze the information in real-time. As a result, the robot can make instant decisions such as avoiding obstacles or fine-tuning machinery settings.

One of the primary technologies enabling real-time data processing in humanoid robots is edge computing. Edge computing moves data processing closer to the data source, reducing latency and bandwidth use. In practical terms, this means that the robot has a small but powerful processor onboard, capable of handling data without needing to send it to a centralized cloud server for analysis. This localized processing is essential for real-time applications where delays of even a few milliseconds could pose safety or efficiency hazards.

Furthermore, real-time data processing isn't an isolated task but is integrated into the broader software architecture of humanoid robots. The operating systems designed for these robots are optimized to handle real-time tasks. These systems are rigorously tested to ensure they can perform under the demanding conditions of a manufacturing environment, where unexpected events are the norm rather than the exception.

Take, for example, a scenario where a humanoid robot works alongside humans on an assembly line. The robot must constantly monitor its surroundings using an array of sensors, ensuring it doesn't collide with a human coworker or another piece of machinery. Through real-time data processing, the robot can dynamically adjust its movements, calculating the optimal path to perform its tasks efficiently and safely. This level of responsiveness is crucial for seamless human-robot collaboration, reducing accidents and improving overall productivity.

Another vital application of real-time data processing in humanoid robots involves predictive maintenance. Traditional maintenance schedules are often based on fixed intervals, regardless of the actual wear and tear on machinery. Predictive maintenance, enabled by real-time data processing, allows robots to continuously monitor the performance and condition of equipment. By analyzing this data, the robots can predict when a machine is likely to fail and prompt preemptive maintenance activities. This approach reduces downtime and extends the lifespan of machinery, leading to significant cost savings.

The impact of real-time data processing extends beyond individual robots to entire networks of humanoid robots working collectively. These robots can share data with each other, learning from each other's experiences and adjusting their behaviors in a coordinated manner. This interconnectedness creates a robust and adaptive production system capable of scaling up or down based on real-time demand and other influencing factors.

Moreover, real-time data processing equips humanoid robots with advanced problem-solving abilities. Suppose a robot encounters an unexpected obstacle or an error on the production line. In that case, it can analyze various data points to determine the best course of action, whether re-routing to avoid the obstacle or isolating the error for a human worker to resolve. This problem-solving capability is crucial for maintaining high operational efficiency and reducing the burden on human workers, enabling them to focus on more complex tasks.

Ensuring the reliability and security of real-time data processing is another crucial challenge. The data gathered and processed by humanoid robots is often sensitive, involving proprietary manufacturing processes or equipment specifications. Thus, robust cybersecurity measures must be in place to protect this data from unauthorized access or malicious attacks. Additionally, ensuring data integrity is essential, as erroneous data could lead to incorrect decisions, potentially harming the production process or even causing accidents.

Several companies and research institutions are at the forefront of developing real-time data processing technologies for humanoid robots. Ongoing innovations in sensor technology, machine learning algorithms, and edge computing are continually enhancing the capabilities of these robots. As these technologies advance, we can expect even greater improvements in the speed and accuracy of real-time data processing, further enhancing the effectiveness of humanoid robots in industrial settings.

Advanced analytics also plays a significant role in leveraging real-time data processing. Humanoid robots can utilize data analytics tools to generate actionable insights from the enormous volumes of data they collect. This not only improves the robots' immediate response capabilities but also contributes to long-term strategic planning. By analyzing patterns and trends, manufacturers can identify areas for improvement, optimize workflows, and make more informed decisions about future investments in technology and infrastructure.

The future holds exciting possibilities for real-time data processing in humanoid robots. One promising avenue is the integration of artificial intelligence with edge AI, enabling robots to learn from real-time data continuously. This could lead to a new era of self-optimizing production, where robots not only react to real-time conditions but also anticipate and prepare for future scenarios based on historical data and predictive algorithms.

In conclusion, real-time data processing is a cornerstone of the software architecture in humanoid robots, enabling them to perform complex tasks with remarkable efficiency and accuracy. This capability revolutionizes industrial environments, driving unprecedented levels of productivity and safety. As technology continues to advance, the potential for real-time data processing in humanoid robots will only grow, paving the way for even more sophisticated and intelligent automation solutions. It's clear that the future of manufacturing will be shaped by the extraordinary capabilities of real-time data processing, providing a competitive edge to industries willing to embrace this transformative technology.

Chapter 6:
The Manufacturing Process: Human vs. Humanoid

In the rapidly evolving landscape of industrial automation, humanoid robots are poised to revolutionize the manufacturing process by offering unparalleled efficiency, precision, and safety. Unlike their human counterparts, these robotic entities can work tirelessly without breaks, execute tasks with consistent accuracy, and operate in hazardous environments without risk. The comparative analysis between human workers and humanoid robots unveils significant improvements in productivity and error reduction, driven by advanced AI algorithms and sensory technologies. However, this transformation isn't just about replacing human labor; it's about augmenting human capabilities and creating a new paradigm where humans and humanoids collaborate harmoniously. This chapter delves into the intricate dynamics of this transition, highlighting how the integration of humanoid robots can lead to safer workspaces and streamlined operations, ultimately reshaping the future of manufacturing.

Comparative Analysis of Efficiency

The advent of humanoid robots powered by AI in manufacturing settings has triggered a surge of interest and debate regarding their efficiency relative to human workers. This shift prompts a detailed examination to assess whether these robots can truly outperform humans in various manufacturing tasks, providing meaningful insights into the comparative efficiency of humanoid robots versus human labor.

At the core of this analysis is the operational speed of humanoid robots compared to human workers. Humanoid robots, equipped with sophisticated AI algorithms and high-speed processors, can execute repetitive tasks with remarkable precision and speed. Unlike their human counterparts, who might need breaks and are prone to fatigue, humanoid robots can operate continuously, ensuring uninterrupted production lines. This capability alone offers a substantial boost in operational efficiency, especially in industries like automotive manufacturing where speed is crucial.

Yet, speed is not the only metric by which efficiency is measured. Accuracy and consistency are equally important. Human workers, despite their skills, may exhibit variations in task performance due to fatigue, stress, or human error. In contrast, humanoid robots offer a consistent level of performance, executing complex tasks with a high degree of precision. They can be programmed to maintain tight tolerances and adhere to exact specifications, thereby reducing the margin of error significantly. This level of consistency translates to fewer defects and a higher quality of end products.

Another critical factor in the efficiency debate is adaptability. Human workers are inherently adaptable; they can switch tasks, learn new skills on the job, and make judgements based on a variety of situational factors. Traditionally, robots have been limited by their programming. However, advancements in AI and machine learning are empowering humanoid robots to become increasingly adaptive. They can now be trained to recognize and adjust to new patterns, anomalies, and changes in their environment. This growing adaptability narrows the gap between human versatility and robotic precision.

When considering operational efficiency, it's also important to evaluate downtime. Human workers are subject to shifts, planned breaks, and unforeseen absences such as sickness or personal emergencies. Conversely, humanoid robots are designed for continuous operation. They don't require breaks and can work around the clock, leading to increased overall productivity. Maintenance and software updates are necessary but can be scheduled during off-peak hours to minimize impact.

However, the integration of humanoid robots also brings about considerations of cost-efficiency. The initial investment in robotic systems can be substantial, encompassing both hardware expenses and the costs associated with programming and implementing these systems. Over time, the cost-benefit ratio of humanoid robots improves as their continuous operation reduces labor expenses, and their efficiency in production processes minimizes wastage and defects.

Energy consumption is another aspect where humanoid robots may have an edge. Modern robots are designed to be energy-efficient, often consuming less power than entire teams of human workers. Additionally, advances in sustainable energy sources and battery technologies further enhance this efficiency. In the long run, the energy savings could contribute significantly to reducing operational costs and environmental impact.

Human workers bring to the table a wealth of creativity and problem-solving skills that robots have yet to fully replicate. While humanoid robots excel in carrying out predetermined, repetitive tasks efficiently, they can struggle with assignments requiring innovative thinking and complex decision-making. These higher-level cognitive tasks are where human workers truly shine, proposing innovative solutions and improvements that robots might miss.

Additionally, the efficiency of humanoid robots can be amplified when they are used in tandem with human workers. Collaborative robots, or cobots, are designed to work alongside humans, augmenting their capabilities rather than replacing them entirely. This symbiotic relationship leverages the best attributes of both robots and humans, leading to a more efficient and harmonious work environment. For example, robots can handle the menial tasks, freeing up human workers to focus on more complex, decision-oriented roles.

Addressing the potential for reduced operational errors, humanoid robots equipped with advanced sensors and real-time data processing systems can detect and respond to issues instantaneously. For instance, if a defect is detected in the manufacturing line, a humanoid robot can halt production and alert human supervisors for further inspection, minimizing wastage and the potential for larger systemic issues down the line.

Communication and coordination are areas where human workers still hold an advantage over robots. The ability to understand subtle cues, adapt to new instructions quickly, and collaborate seamlessly with team members remain domains where humans outperform robots. Even with advances in AI, robots still lack the nuanced understanding and social intelligence that come naturally to humans. The integration of communication and coordination systems into humanoid robots is an ongoing area of development, aiming to bridge this gap.

In summary, while humanoid robots demonstrate significant advantages in speed, precision, and continuous operation, human workers excel in adaptability, creativity, and problem-solving. The comparative efficiency of the manufacturing process largely depends on the specific context and the nature of tasks at hand. Forward-thinking industries are increasingly looking toward hybrid models that leverage the strengths of both humanoid robots and human workers to achieve optimal efficiency.

The continual evolution of AI and robotics promises to further enhance the roles humanoid robots play in manufacturing. Future innovations may bring about even greater efficiency gains, reducing the current limitations of adaptability and communication. As we stand on the brink of this new industrial era, embracing the collaborative potential of humanoid robots and human workers is crucial for realizing the full benefits of technological advancements.

Ultimately, the pursuit of efficiency in manufacturing is a dynamic process, one that requires constant adaptation and improvement. As humanoid robots continue to evolve and their integration into manufacturing processes deepens, ongoing studies and practical experiments will be essential to fully understand and harness their potential. The goal is not merely to replace human workers but to create more efficient, innovative, and sustainable manufacturing environments that benefit from the strengths of both human ingenuity and robotic precision.

Embracing the future of industrial automation involves a recognition of the complementary roles that human workers and humanoid robots can play. Through strategic integration and continuous improvement, we can achieve a manufacturing landscape where efficiency is maximized, errors are minimized, and the potential for innovation is boundless.

Safety and Error Reduction

When discussing the advancements in manufacturing, one cannot ignore the indispensable role of safety and error reduction. The introduction of AI-powered humanoid robots into factory settings marks a pivotal shift toward enhancing operational safety standards and minimizing human error. This transformation, driven by cutting-edge technology, presents a compelling case for integration and broad adoption across diverse industrial landscapes.

Historically, human workers in manufacturing settings have been exposed to various safety hazards. These range from machinery-related accidents to ergonomic injuries caused by repetitive tasks. As humanoid robots step into roles traditionally occupied by humans, they bring a dual advantage: they can perform hazardous tasks without the associated risks and ensure processes are carried out with high precision, thereby reducing error rates.

One of the most notable benefits of humanoid robots is their ability to operate in environments too dangerous for human workers. Whether it's in handling toxic chemicals, working in extreme temperatures, or navigating spaces with heavy machinery, humanoid robots equipped with advanced sensors and AI can perform these tasks with impeccable precision. This not only ensures worker safety but also guarantees the continuity of operations in conditions that would otherwise necessitate shutdowns or slowdowns.

Precision and Repetition: AI-powered humanoids excel in tasks that require repetitive precision, a domain where humans are prone to fatigue and error. Robots don't tire; their efficiency doesn't degrade over time, and they can maintain the same level of accuracy throughout their operation. This translates to a significant reduction in the margin for error—a critical factor in industries where precision is paramount.

The enhanced safety measures brought by humanoids also extend to their collaborative work with human counterparts. Modern humanoids are designed with intricate sensor arrays and sophisticated algorithms that enable them to detect and interpret their surroundings in real-time. This awareness helps prevent collisions and accidents, creating a safer work environment. For instance, if a human inadvertently steps into a robot's path, the robot's sensors can detect this and stop or alter its trajectory immediately, preventing potential injury.

Error reduction is another crucial area where humanoid robots demonstrate their superiority. Traditional manufacturing processes are susceptible to human errors, which can stem from misinterpretation of instructions, lapses in concentration, or physical limitations. Humanoid robots, governed by precise programming and advanced AI, execute tasks flawlessly as per their operational protocols. They rely on exact data, eliminating errors that could arise from human misjudgment or oversight.

Implementation of Predictive Maintenance: One of the fascinating aspects of AI-powered humanoids is their ability to partake in predictive maintenance. Equipped with sensors that monitor machinery health, these robots can identify potential issues before they become severe problems. They continuously analyze data patterns and can alert human supervisors about impending malfunctions. This proactive approach not only reduces downtime by addressing issues early but also minimizes the risk of catastrophic failures that can endanger human lives.

Moreover, humanoid robots incorporate continuous learning mechanisms. Unlike their human counterparts, who might need time to adapt to new practices or rectify their technique, humanoids can instantly update their operational databases with new data. This adaptability means that with each task, they become better at error detection and rectification, further driving down the rates of mishandled operations.

Real-Life Impact: Consider a scenario in an automotive plant where precision in assembling tiny but critical components is imperative. A human worker, after hours of labor, might lose focus, leading to errors that could ripple down the production line. On the contrary, a humanoid robot, utilizing AI algorithms, can perform these delicate tasks with consistent precision, ensuring each component is assembled flawlessly. The result is a significant reduction in defective products, leading to higher overall quality and lower costs associated with rework or recalls.

Training and adaptability also play vital roles in enhancing safety and reducing errors. Humanoid robots can be programmed and reprogrammed for various tasks without the need for extensive downtime. When new processes or safety protocols are introduced, updating an entire fleet of humanoid robots is a matter of simple programming. This flexibility ensures that safety standards evolve without hampering productivity.

In terms of regulatory compliance, humanoid robots offer a distinct advantage. Their operations can be monitored and logged with remarkable detail, ensuring that every aspect of their activity adheres to safety regulations and quality standards. This level of detailed logging provides invaluable data for audits and compliance checks, often surpassing the documentation capabilities of human workers.

Economic Benefits: The economic benefits of enhancing safety and reducing errors through humanoid robots can't be overstated. Fewer workplace injuries translate to lower healthcare costs and compensation claims. Reduced errors lead to higher product quality, fewer returns, and diminished costs related to waste and rework. These benefits culminate in a more straightforward path to achieving regulatory approval for various manufacturing processes and products.

However, the integration of humanoid robots in manufacturing does come with its set of challenges. One notable challenge is the initial investment cost. Companies may find the upfront cost of acquiring and installing humanoid robots prohibitive. Nonetheless, the long-term return on investment, driven by reduced safety incidents and enhanced productivity, makes a compelling case for these investments.

Another challenge lies in the acceptance of humanoid robots by the existing workforce. Workers might fear that their jobs are being threatened, leading to resistance against the new technology. It's vital for companies to address these concerns by illustrating how humanoid robots can take over the hazardous and error-prone tasks, allowing human workers to focus on roles that require cognitive skills and creativity.

Moreover, the interaction between humans and robots must be seamless to ensure that safety and error reduction goals are met. Investing in training programs that educate human workers on how to effectively collaborate with their humanoid counterparts can ease the transition. Such initiatives foster a culture of coexistence, where the strengths of both humans and robots are harnessed for optimal productivity and safety.

In the quest for safer and error-free manufacturing processes, the role of humanoid robots is undeniably transformative. They bring a level of precision, consistency, and safety that is unrivaled by human workers. As technology continues to evolve, these robots are not just becoming smarter but also more adaptable and integrated into our industrial ecosystems.

Future Prospects: Looking ahead, as the integration of AI and advanced sensors in humanoid robots becomes more sophisticated, we can expect even greater strides in safety and error reduction. Future robots may possess enhanced capabilities to predict and adapt to changing conditions in real-time, further minimizing human intervention in hazardous tasks and nearly eliminating errors in manufacturing processes.

In conclusion, the role of AI-powered humanoid robots in ensuring safety and reducing errors in manufacturing processes is both revolutionary and indispensable. By taking over dangerous tasks, reducing human error, and continuously improving through learning and adaptation, humanoid robots are reshaping the landscape of industrial safety and efficiency. The journey toward a safer, more precise manufacturing future is well underway, driven by these remarkable robotic innovations.

Chapter 7:
Collaboration Between
Humans and Robots

In the evolving landscape of industrial automation, the synergy between human workers and AI-powered humanoid robots is proving to be transformative. This collaboration leverages the unique strengths of both entities—combining human creativity, problem-solving, and adaptability with robotic precision, stamina, and data processing capabilities. Optimal human-robot teams are reshaping factory environments by enhancing productivity and reducing error rates. Training programs now focus on equipping workers with the skills to interact and work alongside their robotic counterparts, encouraging a culture of continuous learning and adaptation. As humans teach robots contextual decision-making, robots, in turn, provide real-time feedback and support to enhance human performance. This dynamic partnership is setting the stage for a more efficient and innovative future in manufacturing, where the lines between human intellect and robotic execution blur to create seamless operations.

Optimizing Human-Robot Teams

Collaboration between humans and robots in industrial settings isn't just about placing machines alongside workers; it's about creating symbiotic relationships that enhance productivity and job satisfaction. The essence of optimizing human-robot teams lies in the seamless integration of their unique strengths, making the whole greater than the sum of its parts.

To achieve optimal human-robot collaboration, companies need to focus first on effective communication modalities. Clear and intuitive interaction mechanisms are crucial for ensuring that humans can easily command and understand their robotic counterparts. This includes user-friendly interfaces that employ natural language processing, gesture recognition, and even emotional cues to create a smooth operational experience.

Besides, aligning the roles and responsibilities of humans and robots is key. While robots excel at repetitive, dangerous, and precision-intensive tasks, humans bring in creativity, problem-solving, and nuanced decision-making. Proper role differentiation allows for efficient task allocation and mitigates monotony and risk for human workers. This requires an in-depth analysis of the workflow to allocate tasks that best fit the strengths of each team member, be it human or robot.

Training and education play a pivotal role in this integration. Workers need to be educated not only on how to operate these advanced robots but also on how to collaborate effectively. This is where tailored training programs come into play. These programs should go beyond technical skills and cover areas like team dynamics, communication skills, and conflict resolution. This comprehensive approach ensures that human workers feel competent and confident in their interactions with humanoid robots.

Transparency in operations and decision-making processes is another crucial factor. Humanoid robots are increasingly being outfitted with sophisticated AI algorithms that enable them to make autonomous decisions based on real-time data. For humans to trust and collaborate effectively with robots, it's essential that these decision-making processes are transparent. Real-time data sharing and visual feedback mechanisms can help humans understand and trust robot actions, thus fostering a collaborative environment.

One must not underestimate the importance of fostering an inclusive workplace culture. Robots should not be seen as replacements but as collaborators that augment human capabilities. Building a corporate culture that promotes inclusivity and values the contributions of both humans and robots is crucial for long-term success. This cultural shift can be facilitated through leadership messaging, participative decision-making, and regular team-building activities that include both human and robotic members.

Feedback loops are essential for optimizing human-robot interactions. A continuous cycle of feedback allows for real-time adjustments and improvements in the collaboration process. Regular performance evaluations should consider both human and robotic contributions and provide actionable insights for improvement. Implementing systems for capturing and analyzing this feedback can lead to iterative enhancements in the workflow and collaboration dynamics.

The physical workspace also holds significant importance. Factories need to be reconfigured to facilitate smooth human-robot interactions. Spatial layouts should minimize bottlenecks and ensure that robots have enough room to maneuver without causing disruptions. At the same time, ergonomic considerations for human workers should not be compromised. Creating an environment where both humans and robots can operate safely and efficiently is a critical step in optimizing the team dynamics.

An often-overlooked aspect is the emotional and psychological well-being of human workers. With robots taking over more tasks, humans might experience anxiety about job security or feel isolated in their roles. Companies need to address these concerns proactively by ensuring job roles evolve rather than disappear. This involves redesigning jobs to include more complex and rewarding tasks that humans are uniquely suited for, thereby maintaining their sense of purpose and job satisfaction.

Another key consideration is the adaptability of the teams. In fast-paced industrial environments, flexibility and quick adaptation to changing circumstances are valuable assets. Both robots and human workers need to be able to adjust workflows on the fly. Robots should be equipped with adaptive AI that allows them to learn from their interactions and improve performance over time. Concurrently, humans should be encouraged and trained to develop a mindset of continuous learning and adaptability.

Data security and privacy cannot be ignored when talking about optimizing human-robot teams. As robots collect and process large amounts of data to make real-time decisions, ensuring this data is protected against breaches is crucial. Implementing robust cybersecurity measures will not only help in protecting sensitive information but also in building trust among human workers who might be wary of privacy invasions.

Finally, future-proofing the workforce means investing in ongoing education and development programs. As technology evolves, so do the skill requirements. Continuous learning opportunities, certification programs, and cross-training initiatives can help keep the workforce adept at collaborating with evolving robotic technologies. Organizations should create pathways for career advancement that incorporate these new skills, ensuring long-term workforce resilience.

By focusing on these multifaceted strategies, industries can optimize human-robot teams to not only boost productivity but also create a harmonious and future-ready workplace. The journey involves thoughtful planning, continuous improvement, and a genuine commitment to leveraging the complementary strengths of humans and robots, paving the way for unprecedented innovation and efficiency.

Training and Learning from Each Other

As the collaboration between humans and robots becomes more prevalent in industrial settings, one of the most critical aspects of this collaboration is mutual training and learning. Essentially, this bidirectional flow of knowledge can significantly enhance operational efficiency, safety, and innovation. Humans bring their vast experience, problem-solving abilities, and intuition to the table, while AI-powered humanoid robots offer precision, consistency, and the capacity to handle repetitive, dangerous, or ergonomically challenging tasks.

The first step in building an effective human-robot team is to establish a foundational level of understanding and trust. Hands-on training sessions often serve as the initial phase where human workers get familiar with the robots' capabilities, limitations, and workflows. Such sessions allow humans to understand how to instruct, monitor, and optimize robotic performance. This period of familiarization is crucial to ensuring that humans don't see robots as threats but as collaborative partners designed to make their jobs easier and safer.

Moreover, robots can learn from humans through real-time data capture and machine learning algorithms. When a human operator performs a task, sensors embedded within the robot can record minute details of the operation. This data helps in fine-tuning the robot's algorithms, improving its performance over time. For example, if an operator demonstrates a specialized welding technique, the robot can analyze the steps, speed, and angles involved, thus enhancing its own welding proficiency.

Conversely, humans can also learn from robots. As robots analyze vast amounts of data more efficiently than humans can, they might detect patterns or anomalies that aren't immediately apparent to the human eye. These insights can then be shared with human operators, contributing to continuous learning and improvement. For instance, if a robot identifies a consistent error in assembly line production, it can alert human supervisors who can then adjust procedures or provide additional training to staff.

Learning isn't a one-time event but an ongoing process. Both humans and robots must adapt to new technologies, procedures, and market demands. In this evolving landscape, a framework for continuous learning is essential. Regular workshops, updates, and training sessions ensure that both robots and humans stay current and efficient. Notably, modern factories are increasingly adopting digital twins—virtual replicas of physical robots—to simulate and test new tasks before implementing them on the factory floor. This approach minimizes risks and ensures that both human workers and robots can adapt to changes seamlessly.

Interpersonal skills, surprisingly, also play a role in this collaboration. While robots lack emotions, their design and interaction protocols can significantly impact how they are perceived and accepted by human colleagues. Robots programmed to use polite language, maintain an appropriate distance, and communicate clearly can foster a more harmonious working environment. These seemingly small design choices can make a big difference in how quickly and efficiently human-robot teams can achieve their objectives.

Another facet of training and learning from each other involves troubleshooting and problem-solving. When unplanned issues arise, human ingenuity and robotic precision can synergize to arrive at solutions more quickly. For instance, if a robot encounters an unexpected obstacle or error, it can alert a human operator who can then assess the situation, apply their judgment, and guide the robot toward a resolution. This kind of interplay fosters a more resilient and adaptive manufacturing process.

Additionally, the rise of Augmented Reality (AR) and Virtual Reality (VR) technologies offers novel methods for training and collaboration. AR devices can superimpose digital information over the physical environment, allowing operators to see real-time data and instructions directly within their field of view. VR, on the other hand, can simulate various scenarios for both human and robot training, from emergency procedures to new assembly techniques. These technologies facilitate a more immersive and effective learning experience, bridging the gap between human intuition and robotic precision.

It's also important to note that as robots learn from humans, they must do so ethically and transparently. The data collected during training sessions needs to be handled with care to protect proprietary information and respect privacy concerns. Companies should establish clear guidelines on data usage, storage, and sharing, ensuring that both workers and management are aligned on these protocols.

Despite the clear benefits, integrating this level of collaboration isn't without its challenges. Resistance to change is a natural human tendency that companies must address proactively. Engaging workers in the decision-making process, providing comprehensive training, and highlighting the personal and professional benefits of working alongside robots can ease this transition. Transparency about the role and capabilities of robots can also demystify their presence, transforming them from perceived threats to valued team members.

Interestingly, industries have reported significant morale boosts when robots take over mundane and physically taxing tasks. Workers find that they can focus on more intellectually stimulating work, thus enhancing job satisfaction and productivity. This shift allows employees to leverage their problem-solving skills and creativity, often leading to innovation in other areas of the manufacturing process.

Training programs should also adapt to the varied learning styles of human workers. While some may prefer hands-on learning, others might benefit from visual aids, manuals, or interactive simulations. Tailoring training methods to meet these diverse needs ensures that workers can absorb and apply new information effectively.

Finally, feedback mechanisms should be in place to continuously improve the collaboration between humans and robots. Regular surveys, performance metrics, and even one-on-one discussions can provide insights into what is working well and what needs adjustment. This feedback loop is essential for creating a dynamic environment where both humans and robots can thrive.

In sum, the training and learning period is not merely about imparting knowledge but about creating a symbiotic relationship. When both parties learn from each other, the result is a more efficient, innovative, and harmonious working environment. Through mutual respect, continuous learning, and ethical practices, the collaboration between humans and robots can reach its full potential, heralding a new era in industrial automation.

Chapter 8:
Economic Impacts of
AI-Powered Robots

The introduction of AI-powered robots into industrial settings has irrevocably transformed economic landscapes by optimizing productivity and driving down costs. Companies are witnessing unprecedented efficiencies as these robots take on tasks that were once labor-intensive, leading to reduced operational expenses. Not only do they minimize human error, but they also operate seamlessly around the clock, impacting labor markets and reshaping job roles. While initial investments may appear steep, the long-term economic benefits include increased output, faster turnaround times, and improved product quality. These transformations incentivize industries to adopt AI technologies, leading to a ripple effect of economic growth and innovation across various sectors. Thus, the strategic integration of AI-powered robots is not merely a technological trend but a profound economic shift with far-reaching implications.

Cost-Benefit Analysis for Industries

Integrating AI-powered humanoid robots into industrial settings presents both significant opportunities and challenges. A comprehensive cost-benefit analysis is vital for industries considering this transformation. This analysis will not only highlight direct financial impacts but also delve into operational efficiencies, long-term economic gains, and indirect costs associated with workforce restructuring and technology implementation.

Initial investment costs are a critical factor. AI-powered humanoid robots require substantial upfront capital for both acquisition and installation. This includes the expenses for the robots themselves, the necessary supportive infrastructure such as advanced sensors and networks, and the integration with existing systems. While the cost can be daunting, companies must evaluate this against the long-term savings these robots can generate.

AI-powered humanoid robots excel in performing repetitive, high-precision tasks, thus leading to a marked increase in production efficiency. By operating 24/7 without the need for breaks, these robots can substantially boost output while maintaining consistent quality. This level of reliability often translates into fewer production errors, reducing the costs associated with waste and rework. Consequently, the efficiency gains can offset the initial capital expenditure over time.

The reduction in operational costs is another substantial benefit. Humanoid robots can take over hazardous tasks, minimizing risks to human workers and reducing insurance premiums and other safety-related expenses. Additionally, their ability to operate in environments that would typically wear down human workers can lead to fewer occupational injuries and related costs. Over time, these factors can lead to notable savings on workforce expenses and increase overall safety and well-being within the workplace.

However, the integration of advanced robotics isn't without its challenges. Training and maintenance costs are significant considerations. Skilled technicians and engineers are required to maintain and program these sophisticated machines. This might necessitate the hiring of new staff or investing in substantial training programs for existing employees. Furthermore, regular maintenance and updates are crucial to ensure optimal performance, which represents an ongoing cost.

On the technological front, AI-powered humanoid robots are designed to continuously learn and adapt. This adaptability means that they can be programmed to take on new tasks, providing industries with flexibility. As production demands evolve, the cost of reprogramming or updating a robot is generally lower and faster than retraining a human workforce. This adaptability is a long-term economic advantage, as it allows companies to pivot more quickly in response to market changes.

From an economic perspective, the benefits to industries extend beyond the factory floor. Improved production efficiency and quality can lead to greater customer satisfaction and retention. As robots take over mundane and repetitive tasks, human workers can be redeployed to more creative and strategic roles. This shift can enhance job satisfaction and foster innovation, leading to a more dynamic and competitive business environment.

Additionally, the use of humanoid robots can present new marketing opportunities. Companies can position themselves as leaders in technological innovation, enhancing their brand reputation and attracting new customers and investors. This competitive edge can be particularly advantageous in markets where technological prowess is highly valued.

The financial implications of workforce restructuring cannot be overlooked. As robots take over certain tasks, there may be an initial reduction in the need for human labor. This transition can lead to short-term job displacement and necessitate reskilling and upskilling initiatives to help workers adapt to new roles. While these initiatives represent an additional cost, the long-term benefits of a more skilled and adaptable workforce can outweigh the initial expenses.

Tax incentives and subsidies can further offset the costs of implementing AI-powered robots. Many governments recognize the long-term economic benefits of technological advancement and offer financial incentives to companies that invest in automation. These incentives can significantly lower the net cost of robot integration, making the investment more attractive to industry players.

Another significant aspect of the cost-benefit analysis is the lifespan and durability of AI-powered humanoid robots. With advancements in technology, these robots are becoming more robust and reliable, leading to longer operational lifespans and reduced frequency of replacements. This longevity contributes to a more favorable return on investment (ROI) over time.

Taking a holistic view, the environmental benefits of AI-powered robots also play a crucial role in the cost-benefit analysis. Robots operate with high energy efficiency and can contribute to greener manufacturing processes. Reduced waste, optimized resource usage, and lower emissions align with global sustainability goals, offering an indirect yet significant economic benefit by potentially reducing environmental compliance costs and enhancing corporate social responsibility profiles.

Lastly, as industries adopt AI-powered humanoid robots, it's important to consider the broader economic impact. Increased productivity and efficiency can contribute to economic growth, improved competitiveness in the global market, and the potential creation of entirely new industries and job roles. This broader economic outlook highlights how initial costs can be outweighed by significant long-term economic benefits, both for individual companies and the economy as a whole.

In summary, the cost-benefit analysis for integrating AI-powered humanoid robots into industrial settings reveals a complex interplay of initial investments, operational efficiencies, and long-term economic gains. While challenges remain, particularly in terms of workforce restructuring and ongoing maintenance costs, the potential benefits in productivity, safety, and flexibility offer a compelling case for their adoption. Industries poised to leverage these advanced technologies can expect not only to enhance their operational capabilities but also to gain a competitive edge in an increasingly technology-driven market.

Long-Term Economic Projections

The long-term economic projections of AI-powered robots are shaping up to be more transformative than any technological innovation we've seen in recent history. As AI continues to integrate with robotics, the convergence is setting new benchmarks for efficiency, productivity, and workplace dynamics. So, what can we really anticipate over the next few decades?

In essence, the adoption of AI-powered humanoid robots is expected to significantly boost productivity across multiple industries. This will start a new industrial renaissance, driving economic growth comparable to the inventions of mechanized assembly lines or the internet. It's not just about automating repetitive tasks but also about enabling processes that were previously impossible or impractical because of human limitations.

Economists predict a substantial increase in gross domestic product (GDP) in countries that are early adopters of AI-powered robotics. For example, a McKinsey Global Institute study estimates that AI could add up to $13 trillion to the global economy by 2030, with a significant portion of this stemming from advancements in industrial robotics. This leap in economic value will translate into industries being more competitive on a global scale, allowing them to produce goods more efficiently and at lower costs.

An integral factor to consider is job displacement versus job creation. While the initial implementation of AI robotics is likely to lead to job losses in low-skill and repetitive task segments, new job categories will emerge. Roles that focus on robot maintenance, AI oversight, and advanced analytics interpretation will see increased demand. Moreover, with increased productivity, companies may see revenue growth, enabling them to invest in new ventures and consequently create new jobs.

The longevity of these economic gains will also depend on widespread education and retraining programs. Equipping the workforce with the skills needed to navigate this new landscape is paramount. Governments and private enterprises must collaborate to create robust training programs aimed at reskilling employees who are most at risk of displacement. This proactive approach could soften the impact of job losses and facilitate a smoother transition to a robot-centric workplace.

Labor costs are one of the most significant expenses for any manufacturing operation, and AI-powered robots offer the potential to dramatically reduce these costs. While initial setup and integration are capital-intensive, the long-term savings are substantial. Robots don't need salaries, benefits, or breaks, making them an incredibly attractive investment. These cost savings can be reinvested into growth and innovation, perpetuating a cycle of economic advancement.

A shift to robotics doesn't only mean a change in factories. The ripple effects will be felt in supply chains, logistics, and beyond. AI-driven robots can drastically reduce downtime and increase throughput in manufacturing processes. High efficiency in production lines means quicker turnarounds and more agile supply chains, resulting in faster delivery times and the ability to better meet consumer demands. In turn, this efficiency can enhance customer satisfaction and drive further economic benefits.

Moreover, decreased production costs could lead to lower prices for consumers, expanding the market for various products and services. Affordability tends to drive demand, creating a positive feedback loop where increased production leads to increased consumption, thereby fostering robust economic growth.

Forecasts also suggest that AI-powered robots will play a crucial role in mitigating risks associated with labor shortages. Many advanced economies are grappling with aging workforces and declining birth rates. Robots can fill the gap left by retiring workers, ensuring that industries continue to operate at optimal capacity. This shift can help maintain economic stability and growth, even in the face of demographic challenges.

The global landscape will also see shifts. Countries leading in AI and robotics will gain a competitive edge, potentially shifting economic power balances. Emerging economies investing in these technologies early could leapfrog traditional economic powers, creating a new dynamic in global trade and cooperation. Conversely, countries that lag in adopting AI-powered robots may find themselves at a significant disadvantage, struggling to keep pace with more advanced competitors.

In terms of capital markets, investments in robotics and AI technologies are expected to yield substantial returns. Venture capital is increasingly flowing into startups and established companies focusing on these areas, reflecting strong market confidence in the long-term profitability of AI-powered robotics. This surge in investment will likely spur innovation, driving advancements that could have exponential impacts on economic productivity and growth.

Additionally, property and infrastructure development could see changes. Factory layouts and design will evolve to accommodate robotic systems, potentially revitalizing industrial zones and creating smart manufacturing hubs. These hubs could become magnets for high-tech talent, fostering ecosystems that blend AI research, education, and practical applications.

Moreover, the expected economic benefits will go beyond mere financial metrics. Improved working conditions, reduction of hazardous tasks, and increases in workplace safety are crucial advantages. The declining need for human involvement in perilous work environments will reduce workplace injuries and related costs, decisively impacting both company profits and employee well-being.

One potential downside to consider is the initial phase of economic disparity that may arise. Regions and industries slower to adopt AI-powered robots could experience economic stagnation or decline. Hence, it's vital for policymakers to push for equitable technological diffusion and ensure various sectors and populations aren't left behind in this transformative era.

As we project further into the future, it's exciting to envision a world where AI-powered robots contribute to sustainable development. Efficiency gains from robots could lead to more resources being allocated to research in sustainable practices and green technologies. Imagine factories powered by renewable energy sources, optimized by AI for minimal waste and maximum efficiency. The synergistic effects of AI and green tech could herald an era of sustainable industrial growth.

Let's also not forget the cultural and societal changes that will accompany these economic transformations. With robots picking up the slack in menial jobs, humans will have more time to engage in creative, educational, and recreational activities. This shift could elevate the quality of life, fostering a more enlightened and creative society.

In summation, the long-term economic projections for AI-powered robots are overwhelmingly positive. From boosting GDP and productivity to creating new job roles and fostering sustainable practices, the potential benefits are vast. However, realizing these benefits will require coordinated efforts in education, policy-making, and investment. The future indeed looks promising, but proactive measures will be critical in navigating the complexities that come with such monumental change.

Chapter 9:
Challenges and Solutions

Navigating the realm of AI-powered humanoid robots in industrial automation comes with its share of formidable challenges and innovative solutions. Technical limitations, such as the complexity of machine learning algorithms and real-time data processing, often stymie progress. Nevertheless, advancements in computational power and sensor technologies are steadily pushing these boundaries. Another significant hurdle lies in workforce restructuring, as the integration of humanoid robots necessitates a shift in job roles and skill requirements. Forward-thinking companies are investing in retraining programs to equip employees with the necessary skills to collaborate effectively with their robotic counterparts. By addressing these challenges head-on, industries can harness the full potential of humanoid robots, setting the stage for a harmonious and productive future where humans and machines work in unison.

Overcoming Technical Limitations

The journey of integrating AI-powered humanoid robots into factory environments has not been without its hurdles. Despite advancements in robotics and artificial intelligence, numerous technical limitations still challenge widespread adoption. Addressing these limitations is crucial for unlocking the full potential of these technologies in industrial settings.

One of the primary technical limitations revolves around computational power. AI algorithms for humanoid robots require substantial processing capabilities to perform complex tasks in real-time. Traditional processors often struggle to handle the massive data influx, intricate calculations, and rapid decision-making needed for these robots to function effectively. The advent of specialized AI chips and processors designed for high-efficiency computing marks a significant step towards mitigating this issue. These technological advances ensure that robots can perform basic and advanced functions with greater speed and accuracy.

Moreover, the intricate design and complex programming required to give humanoid robots human-like mobility and dexterity present further challenges. Developing actuators and sensors capable of mimicking human movements while maintaining balance and stability is no small feat. Engineers and researchers have been working on enhancing robotic limbs, joints, and motor functions to approximate the fluidity and versatility of human actions.

Power consumption and battery life are also significant technical constraints. AI-powered humanoid robots often require constant recharging or must be tethered to a power source, limiting their operational efficiency. Innovations in energy storage, battery technology, and dynamic power management are vital to overcoming this obstacle. Future developments in energy-efficient hardware components could extend battery life and reduce the overall power consumption of these robots.

Beyond hardware, software limitations pose another critical area needing attention. Developing robust and reliable software architectures capable of handling real-time data processing and decision-making is imperative. As robots operate in dynamic and unpredictable factory environments, their software systems must be adaptive and resilient, capable of seamless integration with existing enterprise systems. This requires an amalgamation of cutting-edge software practices, real-time operating systems, and robust network infrastructures to facilitate uninterrupted communication and data flow.

Interoperability between various systems and robots also presents a notable challenge. Factories often employ a diverse array of machines and sensors from multiple vendors, each with its proprietary technology. Achieving seamless communication between these heterogeneous systems demands the development of standardized protocols and interfaces. Open standards and collaborative partnerships between technology providers could play a crucial role in bridging these gaps, ensuring that humanoid robots can work harmoniously within a broader industrial ecosystem.

Another significant technical limitation is the accuracy and reliability of sensory data interpretation. AI-powered humanoid robots rely heavily on sensors to perceive their environment. These sensors must not only gather data but also interpret it accurately to inform decision-making processes. Improved sensor technologies and advanced data interpretation algorithms are essential for enabling robots to operate with high precision in complex industrial environments.

The integration of machine learning models for object recognition, navigation, and task execution further complicates the landscape. Machine learning algorithms often require extensive training on diverse datasets to achieve high levels of accuracy and reliability. Ensuring that these models can perform consistently across different scenarios and factory layouts requires ongoing calibration and optimization.

Additionally, the sheer volume of data generated by AI-powered robots necessitates robust data management systems. Factories must collect, store, and analyze vast amounts of information to facilitate real-time decision-making and predictive maintenance. Cloud computing and edge AI offer promising solutions, providing the necessary computational resources and storage capabilities. However, ensuring data security and low-latency communication remains a priority to capitalize fully on these technologies.

Furthermore, sensory limitations can impact a robot's ability to function in complex environments. Enhanced sensor technology, including advancements in visual recognition, tactile feedback, and spatial awareness, is necessary to provide humanoid robots with the situational awareness required for intricate tasks. This development encompasses sensors that can detect minute environmental changes, recognize objects accurately, and process input from various modalities simultaneously.

Importantly, there's an ongoing need for user-friendly interfaces that facilitate interaction between humans and robots. Intuitive control systems, programming environments, and communication protocols must be developed to enable factory workers to seamlessly operate, supervise, and collaborate with humanoid robots. User-centric design principles should guide the development of these interfaces, ensuring that they cater to the needs and capabilities of the workforce.

One can't overlook the importance of a resilient and scalable network infrastructure to support the advanced functionalities of humanoid robots. High-speed, low-latency networks are essential to ensure timely data transmission and coordination between different robotic units. The deployment of 5G technology and advanced wireless communication systems can significantly enhance the performance and reliability of these robotic systems in an industrial setting.

Lastly, the future will likely see continuous improvement in AI and robotic technologies through iterative development and practical deployments. Pilot programs, real-world testing, and continuous feedback loops will enable engineers and researchers to refine and optimize these systems. Collaborative efforts between academia, industry, and regulatory bodies will be essential in navigating the complexities and addressing the evolving technical challenges.

In conclusion, while the integration of AI-powered humanoid robots into factories is fraught with technical limitations, the relentless pursuit of innovative solutions is gradually overcoming these hurdles. Combining advancements in hardware, software, and networking with ongoing research and development efforts will pave the way for a new era of industrial automation, significantly enhancing productivity, efficiency, and safety. The continued focus on overcoming these technical challenges promises to unlock the full transformative potential of humanoid robots in industrial settings, ushering in a future where human and robotic collaborations are the norm.

addressing workforce restructuring

The integration of AI-powered humanoid robots in industrial settings marks a transformative phase in manufacturing, promising unprecedented improvements in efficiency and productivity. However, this seismic shift brings with it significant challenges, notably workforce restructuring. Addressing these challenges requires a multifaceted approach that combines technological innovation with strategic human resource management.

One of the primary concerns is the displacement of workers. As robots become more adept at performing tasks that once required human intervention, there is an inevitable reduction in the need for human labor in certain roles. This can lead to job losses, especially in areas where repetitive or dangerous tasks are prevalent. To mitigate this, companies must invest in retraining programs to upskill their employees, preparing them for new roles that leverage human ingenuity and adaptability.

Retraining initiatives should focus on developing skills that are complementary to robotic automation. For instance, workers can be trained in robot maintenance, programming, and oversight roles that ensure the smooth operation and continuous improvement of robotic systems. By shifting the workforce towards more technical and supervisory positions, companies can not only preserve jobs but also enhance the overall productivity and innovation within the factory.

Moreover, there's a significant opportunity for creating new job categories that didn't previously exist. The advent of AI-powered humanoids necessitates roles such as robot behavior analysts, AI trainers, and human-robot interaction specialists. These positions capitalize on human creativity, critical thinking, and problem-solving abilities, areas where humans still hold a distinct advantage over machines. Organizations must be proactive in identifying these emerging roles and providing the necessary educational pathways to fill them.

On a broader scale, there needs to be a collaboration between industries, educational institutions, and governments to create a supportive ecosystem for workforce restructuring. Policies and incentives that encourage continuous learning and adaptability can help ease the transition. Educational curricula must be updated to include subjects like AI, robotics, and data science from an early age, preparing future generations for a technologically advanced workforce.

The psychological impact of workforce restructuring cannot be ignored either. The fear of job loss and the uncertainty of future roles can lead to increased stress and reduced morale among workers. It's crucial for companies to maintain transparent communication with their employees, explaining the long-term benefits of automation and how it aligns with the company's vision. By involving employees in the transition process and seeking their input, organizations can foster a culture of inclusivity and collaboration.

In addition to internal restructuring, the rise of AI-powered humanoid robots can lead to wider economic implications. Smaller businesses and industries dependent on manual labor may struggle to compete with larger firms adopting advanced robotics. This disparity could widen economic inequality, creating a pressing need for policies that ensure equitable access to technology and support for small and medium enterprises (SMEs).

Automation does not necessarily equate to a reduction in the workforce; rather, it can be seen as an opportunity to reallocate human resources towards more strategic and creative endeavors. Many tasks still require the nuanced judgment and emotional intelligence that only humans can provide. For instance, customer service roles, jobs that require empathetic interactions, and complex decision-making processes cannot be easily replaced by robots. Integrating AI-powered robots can free up human workers to focus on these high-value tasks, thereby enhancing overall job satisfaction and productivity.

The role of leadership is pivotal in navigating the challenges of workforce restructuring. Leaders must possess a forward-thinking mindset, embracing technological advancements while being empathetic to their workforce's concerns. Compassionate leadership that prioritizes employee well-being can significantly ease the transition and ensure that automation benefits both the company and its employees.

The successful implementation of workforce restructuring strategies requires meticulous planning and continuous evaluation. Metrics for assessing the impact of automation on employment levels, productivity, and employee satisfaction should be established and monitored. Continuous feedback loops will allow organizations to make data-driven decisions and adjust their strategies as necessary to achieve optimal outcomes.

Regulatory frameworks will also play a significant role in ensuring a balanced transition. Governments must develop and enforce policies that protect workers' rights in the age of automation while encouraging innovation and adoption of new technologies. This includes regulations around fair wages, job security, and ethical considerations in the deployment of AI and robotics.

An often-overlooked aspect is the cultural shift that accompanies the introduction of humanoid robots. Companies need to cultivate a workplace culture that embraces innovation and change. This involves not only technical training but also fostering an environment where employees feel valued and part of the technological evolution. Encouraging a mindset of lifelong learning and curiosity will be essential as the pace of technological advancement shows no signs of slowing down.

Finally, the global nature of industrial manufacturing means that workforce restructuring strategies must consider international contexts. Different regions face unique challenges and opportunities when it comes to automation. Understanding and adapting to these regional variations will be crucial for multinational corporations aiming to implement cohesive and effective workforce management strategies across their global operations.

Addressing workforce restructuring in the face of AI-powered humanoid robots is no small feat. It demands a holistic approach that combines technological innovation, strategic human resource management, continuous education, supportive policies, and empathetic leadership. By tackling these challenges head-on, we can ensure a future where both humans and robots coexist harmoniously, each playing to their strengths and driving industrial progress.

Chapter 10:
Ethical and Societal Implications

As AI-powered humanoid robots become more ingrained in factory environments, their ethical and societal implications can't be overlooked. These robots challenge traditional notions of labor, raise questions about job displacement, and necessitate new frameworks for public perception and acceptance. It's crucial to consider how these machines will coexist with human workers, ensuring that ethical considerations are at the forefront of their deployment. Balancing efficiency gains with societal impacts requires a thoughtful approach where benefits don't come at a societal cost. As we venture into this new era of automation, fostering open dialogues will be key to a harmonious integration of humanoids in our workplaces.

Public Perception and Acceptance

The integration of AI-powered humanoid robots in industrial environments represents a monumental shift in the manufacturing landscape. However, such a significant transformation does not come without public scrutiny and varying levels of acceptance. As organizations push the boundaries of technological capabilities, public perception becomes a critical factor that influences the pace and scope of adoption. Understanding public sentiment is essential for any industry looking to capitalize on the myriad benefits that come with AI-powered humanoid robots.

Initially, the concept of humanoid robots was met with a mix of wonder and skepticism. The general public often associates robots with popular culture, leading to both fascination and apprehension. Media portrayals, whether optimistic or dystopian, have a profound impact on how people perceive robotic technology. For instance, movies and television shows that depict robots as sentient beings capable of surpassing human intelligence can stoke fears regarding job displacement and ethical dilemmas. Conversely, positive representations can foster excitement and acceptance.

The practical implications of AI-powered humanoid robots in the workplace further complicate public perception. On one hand, the promise of enhanced efficiency, reduced error rates, and improved safety regulations are tangible benefits that industries can tout. On the other hand, concerns about job displacement and economic inequality must be addressed thoughtfully. Workforce restructuring is a sensitive topic, especially in regions where manufacturing jobs serve as a crucial economic backbone.

Surveys and studies consistently show a polarized view of robotic integration among workers. Younger employees, more familiar with digital technologies, tend to be more accepting of humanoid robots. They often view these robots as collaborative tools that can augment their work experience. Older workers, however, may feel threatened by AI technologies, fearing that their skills may become obsolete. This dichotomy underscores the importance of comprehensive training programs and educational initiatives aimed at facilitating smooth transitions and engendering a collaborative environment.

Community outreach and transparent communication contribute significantly to public acceptance. Companies that invest in educating the public about the roles and capabilities of humanoid robots often find it easier to achieve acceptance. Public demonstrations, factory tours, and open forums enable direct interaction between the robots and the community, effectively demystifying the technology. Such initiatives can help shift public perception from skepticism to informed optimism.

Public policies and regulatory frameworks also play a pivotal role in shaping acceptance. Governments can ease public concerns by establishing clear guidelines on the ethical deployment of AI-powered humanoid robots. Regulations that emphasize the protection of worker rights and outline the gradual integration of AI technologies help build trust. A regulatory environment that addresses safety, accountability, and ethical considerations is more likely to win public approval.

Moreover, collaborations between academic institutions, industry leaders, and governmental bodies can create a holistic approach to workforce transitions. Educational programs tailored to emerging technologies and vocational training aimed at upskilling the current workforce can mitigate fears of job loss. In the long run, these programs contribute to a more adaptive and resilient labor market, better prepared to embrace technological advancements.

The influence of social media can't be overlooked in shaping public perception. Platforms like Twitter, Facebook, and LinkedIn serve as battlegrounds for opinions on technological progress. Companies must engage in proactive social media strategies to manage narratives and counteract misinformation. By providing accurate, timely information and showcasing real-world success stories, businesses can build a more informed and accepting public.

Understanding global perspectives is equally critical. Different cultures and economies have varying levels of technological acceptance. In countries with a strong focus on innovation and technology, the assimilation of humanoid robots may meet less resistance compared to regions where traditional manufacturing practices are deeply entrenched. Global companies must therefore tailor their strategies to align with regional sentiments and cultural nuances.

One intriguing aspect is the psychological influence of humanoid robot design on public acceptance. Humanoid robots that display human-like features and gestures are often more readily accepted by people. This phenomenon, known as the "uncanny valley," suggests that robots need to balance human-like attributes without appearing too lifelike, which can be unsettling. Designing robots that evoke a sense of familiarity and approachability can help garner public support.

Public perception also hinges on ethical considerations and societal values. As AI-powered humanoid robots become more prevalent, questions surrounding privacy, data security, and the ethical treatment of robots themselves gain prominence. Industries must navigate these ethical waters carefully to gain public trust. Transparent policies on data usage and clear ethical frameworks can assuage public concerns and foster a sense of responsible innovation.

The potential societal benefits of humanoid robots cannot be overlooked. From addressing labor shortages to enabling more efficient and safer work environments, the advantages are numerous. Highlighting these benefits and real-world applications can shift public perception positively. Case studies illustrating successful implementations and tangible improvements in workplace conditions can serve as powerful endorsements.

Ultimately, achieving widespread acceptance of AI-powered humanoid robots in industrial settings requires a multifaceted approach. Balancing technological advancements with ethical considerations, prioritizing educational initiatives, and engaging in transparent communication are key strategies. By addressing public concerns thoughtfully and proactively, industries can pave the way for a future where humanoid robots are not just accepted but embraced as indispensable partners in industrial production.

Ethical Considerations in Deployment

The rise of AI-powered humanoid robots in industrial environments brings with it a cascade of ethical questions that need careful consideration. As we venture into an era where robots will be ubiquitous in manufacturing, it's important to weigh the benefits against potential ethical pitfalls. The deployment of these sophisticated machines isn't merely a technical challenge; it encompasses a range of societal and moral dimensions that must be addressed to ensure their responsible integration.

First and foremost, one of the paramount concerns is the impact on employment. While automation and robotics can significantly enhance productivity and efficiency, there is a legitimate fear that they could displace human workers. Job displacement can lead to economic instability for individuals and communities. As we push forward with the deployment of humanoid robots, it's crucial to develop strategies that include re-training programs, support for displaced workers, and policies that encourage upskilling of the current workforce.

Furthermore, the introduction of humanoid robots in factories raises questions about human dignity and the value of human labor. What is the intrinsic worth of human work when robots can perform the same tasks, often with greater precision and without fatigue? This issue touches on deeper philosophical questions about the role of humans in the workforce and how societies value different forms of labor.

Privacy is another significant ethical consideration. Humanoid robots equipped with advanced sensors and data collection capabilities can gather vast amounts of information about their environment and the people within it. This data collection can lead to concerns about surveillance and the potential for misuse of personal information. It's essential to establish robust data protection standards and ensure transparency in how data collected by robots is used and stored. Factories must strike a balance between leveraging the benefits of data for improving efficiency and maintaining the privacy rights of individuals.

The transparency and accountability of AI decision-making also come under scrutiny. As humanoid robots become more autonomous, the need for clear, understandable, and transparent AI algorithms becomes imperative. Workers and stakeholders need to trust that the systems in place are fair and unbiased. This involves rigorous testing, validation, and the establishment of clear protocols for decision-making processes embedded in these robots.

Moreover, safety is a non-negotiable aspect of ethical deployment. Robots working alongside humans must adhere to the highest safety standards to prevent accidents and injuries. This necessitates not only robust physical design but also sophisticated AI that can predict and avoid potentially hazardous situations. Regulations must be stringent, and constant monitoring should be in place to ensure that robots operate within safe parameters at all times.

There's also the matter of consent and the relationship between humans and robots. Workers interacting with robots should have a clear understanding of what these machines are capable of and what limitations they have. Informed consent becomes essential, particularly in scenarios where robots might make autonomous decisions that impact human workers. This requires comprehensive training and clear communication from employers.

Another layer of ethical consideration is the cultural and societal acceptance of robots. Different regions and communities might have varying levels of comfort and acceptance of humanoid robots. Cultural norms and values can heavily influence how robot deployment is perceived. Ethical deployment strategies should be culturally sensitive and consider local contexts, ensuring that the introduction of robots doesn't create friction or resistance within the workforce and the wider community.

Inclusion and accessibility also play a critical role. It's important to ensure that the integration of humanoid robots does not exclude or marginalize individuals with disabilities or those who might find it challenging to adapt to new technologies. Humanoid robots should be designed with universal design principles in mind, making them accessible to all.

Ethical considerations also extend to the environmental impact of deploying humanoid robots. Sustainable practices in manufacturing these robots, as well as their energy consumption and lifecycle management, should be prioritized to mitigate negative ecological impacts. Ethical deployment involves not just maximizing efficiency and productivity, but also ensuring that these advancements align with sustainable development goals.

We must also address the long-term societal implications. The rise of humanoid robots in factories can influence societal structures, economic models, and even global trade dynamics. Policymakers, industry leaders, and society at large must engage in continuous dialogue to navigate these changes in a way that benefits the collective good.

Ultimately, the journey toward integrating AI-powered humanoid robots into industrial settings is fraught with complex ethical challenges. However, by approaching these challenges with a grounded, thoughtful perspective, it's possible to harness the immense benefits of robotics while mitigating potential downsides. This requires a multifaceted approach, involving policy formulation, regulatory frameworks, community engagement, and a steadfast commitment to ethical principles.

The deployment of humanoid robots isn't merely a technological evolution; it's a societal transformation. By acknowledging and addressing the ethical considerations, we can ensure that this transformation is both beneficial and equitable, paving the way for a future where humans and robots collaboratively enrich the fabric of industrial production.

Chapter 11:
Case Studies of Successful Implementations

In this chapter, we delve into the remarkable successes encountered by leading companies that have wholeheartedly embraced AI-powered humanoid robots in their industrial processes. Highlighting diverse examples from a variety of sectors, this chapter uncovers how these businesses have harnessed advanced robotics to elevate their operational efficiency, quality, and agility. Through detailed case studies, we will illustrate the tangible benefits these pioneers have achieved, including reduced downtime, enhanced precision, and significant cost savings. The insights gleaned from early adopters serve not only as a roadmap for future implementations but also as a testament to the transformative potential of humanoid robots in the modern industrial landscape. Lessons learned from these trailblazers emphasize the importance of strategic planning, continuous adaptation, and the cultivation of a collaborative human-robot workforce.

Leading Companies Embracing Humanoids

As humanoid robots gain traction in industrial environments, several forward-thinking companies are leading the charge. These companies have not only recognized the potential of humanoids to revolutionize factory floors but have also implemented these advanced robots in ways that optimize both efficiency and safety. This section will highlight some of these industry pioneers and the innovative ways they're employing humanoid robots.

One such company is *Toyota*. With its long history of adopting cutting-edge technology, it's no surprise that Toyota is at the forefront of humanoid robot integration. The car manufacturer utilizes a range of humanoid robots in its production lines to perform tasks such as welding, assembly, and quality inspection. These robots, equipped with advanced AI algorithms and sensor technologies, can adapt to various tasks and conditions, providing flexibility beyond traditional robotics. This adaptability significantly reduces production downtime and increases overall efficiency.

Boston Dynamics, known for its groundbreaking work in robotics, has also made significant strides in developing humanoid robots for industrial applications. Their humanoid robot, Atlas, has been employed in several factories for its exceptional mobility and dexterity. Using machine learning, Atlas can navigate complex environments and perform tasks that were previously reserved for human workers. The robot's human-like abilities enable it to interact seamlessly with human teams, boosting the overall productivity and versatility of manufacturing processes.

In the electronics sector, *Foxconn* has recognized the value of humanoid robots in enhancing productivity. These robots are particularly useful in assembly lines where precision and speed are paramount. Foxconn's humanoid robots handle delicate components and perform repetitive tasks with a level of consistency and accuracy that is difficult to achieve with human labor alone. This not only accelerates the production process but also significantly reduces the margin of error, which is crucial in electronics manufacturing.

Meanwhile, **Siemens** has integrated humanoid robots into its smart factories initiative. In these environments, humanoid robots work alongside human operators, taking on more strenuous and hazardous tasks. By doing so, Siemens has managed to create a safer work environment while maintaining high efficiency and productivity levels. The robots' ability to learn and improve from interactions ensures continuous optimization of manufacturing processes.

ABB, another leader in automation technologies, has been leveraging humanoid robots for years. Their YuMi robot, initially designed for collaborative tasks, demonstrates how humanoids can work hand-in-hand with human operators. ABB has taken this concept further by integrating AI to enhance YuMi's capabilities, allowing it to take on more complex and varied tasks. The ongoing implementation of humanoid robots at ABB's facilities showcases their commitment to innovation and efficiency in manufacturing.

In the aerospace industry, **Airbus** is utilizing humanoid robots to meet the high standards required in aircraft manufacturing. These robots are employed in tasks ranging from component assembly to quality checks, where precision is paramount. The integration of humanoid robots helps Airbus maintain stringent quality control while speeding up production timelines, crucial for meeting delivery deadlines in the aerospace sector.

Amazon has also embraced humanoid robots in its fulfillment centers. These robots are key in optimizing logistics and inventory management. By automating the handling of packages and products, Amazon's humanoid robots alleviate the physical strain on human workers and ensure quicker processing times, enhancing overall operational efficiency.

Moreover, **Samsung** is leveraging humanoid robots in its manufacturing processes. These robots are especially beneficial in tasks that require precision and consistency, such as semiconductor manufacturing. The ability to perform these tasks without human intervention not only speeds up the production cycle but also ensures a higher level of precision, critical in the semiconductor industry.

From these examples, it's evident that various industries are progressively adopting humanoid robots to enhance their operations. Whether it's automotive, electronics, aerospace, or logistics, the integration of humanoid robots presents significant benefits in terms of efficiency, safety, and productivity. These companies exemplify how strategic implementation of advanced robotics can revolutionize industrial environments.

Furthermore, by embracing humanoid robots, these leading companies are setting a precedent for others to follow. They showcase the potential of AI-powered robotics in transforming traditional manufacturing processes, paving the way for a future where robots and humans work synergistically. This collaborative approach not only propels technological advancement but also fosters a culture of continuous improvement and innovation.

In conclusion, the move towards incorporating humanoid robots is not just a trend but a strategic evolution in manufacturing and industrial operations. As these leading companies continue to innovate and optimize their use of humanoid robots, the broader industry will undoubtedly feel the ripple effects, leading to more widespread adoption and continued advancements in the field of robotics.

Lessons Learned from Early Adopters

Early adopters of AI-powered humanoid robots in industrial settings have provided invaluable insights into the practical application of this ground-breaking technology. These pioneers have paved the way for broader adoption, facing challenges head-on and developing innovative solutions that set a precedent for those who follow. Through their experiences, we've gleaned lessons on implementation strategies, the importance of adaptability, and the unforeseen benefits and issues that arise when integrating advanced robotics into traditional manufacturing processes.

One of the principal lessons learned is the importance of initial assessment and planning. Early adopters quickly realized that successful integration required more than just purchasing robots and installing them on assembly lines. Extensive groundwork was necessary to evaluate the factory's current workflow, identify the roles suitable for humanoid intervention, and understand how robots can complement rather than compete with the human workforce. Companies that invested time in comprehensive feasibility studies reported fewer setbacks and smoother transitions.

Another significant insight is related to the customization of robots. Humanoid robots, although highly advanced, are not one-size-fits-all solutions. Early adopters discovered that tailoring these machines to the specific needs of their operational environment maximized efficiency. Adjustments in the robots' algorithms, sensor configurations, and even physical modifications were often necessary to align with the unique demands of each manufacturing process.

Moreover, the importance of continuous training and development cannot be overstated. Initial adopters found that setting up a robust training program for employees working alongside humanoid robots was crucial. Not only did this help in reducing resistance and anxiety among staff, but it also optimized the collaboration between human and machine, resulting in higher productivity and safety. Training programs also evolved to include lessons learned from day-to-day operations, creating a feedback loop that continually improved the human-robot interaction.

One area that presented both opportunities and challenges was data management. Humanoid robots generate a tremendous amount of data, and early adopters highlighted the importance of having robust data handling and analysis capabilities. Companies that could effectively leverage this data gained insights that significantly improved operational efficiency and product quality. Conversely, those who were unprepared for the data influx faced bottlenecks and lost opportunities.

Adjusting to these technological advancements required a cultural shift within the organization. This aspect was underlined by several early adopters who emphasized the need for leadership to foster a culture that embraces innovation and change. Companies that succeeded in this regard actively involved their workforce in the transition process, making them a part of the narrative and addressing their concerns comprehensively. This collective ownership over the new technology fostered a more harmonious and effective adoption.

Initial cost is another lesson habitually cited by early adopters. The integration of advanced humanoid robots is no small financial undertaking, and those who underestimated the initial investment found themselves facing unforeseen challenges. However, companies that planned for these costs and considered them as part of a long-term investment saw significant returns. They noted improvements in production speed, accuracy, and overall profitability, proving the substantial long-term benefits outweighed the upfront costs.

Early adopter experiences also showcased the importance of agility and adaptability. Problems often arose that had not been anticipated, requiring companies to pivot and make rapid adjustments. Whether it was a software glitch, a hardware malfunction, or an unexpected operational issue, the ability to respond quickly and adapt strategies was critical for maintaining momentum and achieving success.

Moreover, these initial ventures highlighted the importance of partnerships and collaboration. Many early adopters engaged with experts from academia, technology providers, and other industries to troubleshoot issues and refine the deployment of humanoid robots. These collaborations not only provided technical solutions but also enriched the adopters' understanding of the broader implications and potential of AI-powered robotics.

Safety protocols also emerged as a focal point of learning. Integrating humanoid robots introduced new safety risks that required fresh approaches and updated protocols. Early adopters stressed the importance of developing rigorous safety standards, regular training, and clear guidelines to prevent accidents and ensure a secure working environment. Companies that prioritized safety saw fewer incidents and enjoyed smoother operations.

The psychological aspect of working alongside robots was another lesson underscored by pioneers. The novelty of humanoid robots can be intimidating, and initial interactions often come with apprehensions. Successful early adopters made concerted efforts to humanize the robots, not just in appearance but also in functionality and interaction. This included programming robots to perform non-threatening gestures and incorporate elements of emotional intelligence, which helped workers gradually get accustomed to their new robotic colleagues.

Adaptability in production processes was another critical area of learning. Early adopters frequently had to re-engineer existing workflows to fully exploit the capabilities of humanoid robots. This sometimes involved substantial changes, like redesigning assembly lines or repurposing job roles, illustrating the need for flexibility when embarking on this technological transition.

Finally, one of the most encouraging lessons from early adopters was the potential for innovation. As companies began to understand and utilize their new humanoid partners, they often discovered new ways to improve their products and processes that they hadn't previously considered. Innovation wasn't just a byproduct; it became a driving force for continuous improvement and competitive advantage.

In summary, the early adopters of humanoid robots in industrial settings have provided a wealth of knowledge that future implementers can draw upon. By sharing their experiences, these pioneers have underscored the importance of preparation, customization, training, cultural adaptation, data management, financial planning, agility, collaboration, safety, psychological considerations, process adaptability, and innovation. These lessons form a comprehensive blueprint for anyone looking to integrate AI-powered humanoid robots into their manufacturing processes successfully.

Chapter 12:
Regulatory and Safety Standards

As AI-powered humanoid robots become increasingly integral to manufacturing operations, establishing robust regulatory and safety standards is essential. Regulatory frameworks must evolve to address the unique challenges posed by these advanced machines, ensuring they operate within clearly defined guidelines to protect human workers and maintain operational reliability. Safety standards, often established by organizations such as ISO and OSHA, must account for the intricate interactions between humans and robots in dynamic industrial settings. These standards should cover everything from physical safety measures to cybersecurity protocols, given that AI systems rely heavily on data integrity. The development of these regulations and standards will require a collaborative effort between industry experts, policymakers, and researchers to create a safe and efficient environment for both humans and machines. Clear and enforceable standards will not only bridge the gap between innovation and safety but will also drive public trust and acceptance, paving the way for wider adoption of AI-powered humanoid robots in factories around the globe.

Current Regulations in Industrial Robotics

In today's rapidly evolving industrial landscape, regulations in robotics have become a focal point for ensuring both safety and efficiency. As AI-powered humanoid robots are increasingly being integrated into manufacturing environments, it is vital to understand the current regulatory framework governing their deployment and operation.

The regulatory landscape for industrial robotics is multifaceted, encompassing general safety standards, specialized guidelines for robotics, and emerging regulations tailored for AI-integrated systems. Traditionally, industrial robotics regulations have focused on ensuring the mechanical and electrical safety of robots working alongside humans. Standards like the ISO 10218 set the benchmark for safety requirements in robotics, detailing aspects such as emergency stop functions, safety-rated monitored stops, and manual reset. These guidelines are critical because they help mitigate risks associated with mechanical failures and unintended interactions between humans and robots.

Yet, as robots evolve from simple, repetitive task machines to sophisticated, AI-driven systems, the regulatory requirements are also experiencing a transformation. AI integration introduces a new set of variables, including decision-making capabilities, learning algorithms, and autonomous actions, which traditional safety standards weren't designed to address. This shift has necessitated the development of new regulations aimed specifically at AI-powered industrial robots. The main objective is to ensure that these advanced systems can operate safely and efficiently in complex industrial settings.

One of the key regulatory bodies contributing to the advancement of robotics safety standards is the International Organization for Standardization (ISO). The ISO 15066 standard, for example, offers guidelines on the safe interaction between collaborative robots and human employees. It goes beyond mechanical safety, introducing parameters like pain thresholds and allowable forces to ensure human-robot interactions are both safe and ergonomically designed. This is particularly crucial in environments where robots and humans work in close proximity.

Another noteworthy development is the European Union's AI Act, which aims to create a comprehensive regulatory framework for AI systems, including those used in industrial robotics. This legislation seeks to balance the promotion of innovation with the need to address potential risks associated with AI technologies. It includes provisions for risk assessments, transparency, and accountability, which are essential for fostering public trust and ensuring the responsible deployment of AI-powered robots.

In the United States, the Occupational Safety and Health Administration (OSHA) sets forth guidelines and regulations that impact industrial robotics. Although OSHA's regulations primarily focus on general workplace safety, they also encompass specific standards for automated systems and machinery, placing an emphasis on ensuring that robotic cells are designed and maintained to prevent workplace injuries. Through continuous updates and collaborations with other regulatory bodies, OSHA aims to keep pace with the advancements in robotic technologies and their applications in manufacturing.

Interestingly, regulatory approaches vary globally, reflecting different levels of technological advancement and cultural attitudes toward automation. While Europe and the United States are leading the charge in formalizing AI and robotics regulations, countries in Asia, particularly Japan and South Korea, are also significant contributors to the regulatory discourse. Japan's Ministry of Economy, Trade, and Industry (METI) has issued guidelines focusing on the ethical and societal implications of AI, stressing the importance of trust and coexistence between humans and robots. Meanwhile, South Korea has established the Robot Ethics Charter, which aims to address ethical and safety concerns related to the deployment of robots in various sectors, including industrial manufacturing.

The complexity of AI-powered humanoid robots also introduces challenges related to data privacy and cybersecurity. Regulations are evolving to address these concerns, ensuring that the vast amounts of data processed by these systems are handled securely and ethically. The General Data Protection Regulation (GDPR) in the European Union, though not specific to robotics, has a significant impact on how data is managed within AI systems. This regulation underscores the need for transparency, user consent, and stringent data protection measures, which are critical for maintaining the integrity and trustworthiness of AI-powered robots.

In parallel, industry-specific consortia and professional organizations are actively contributing to the development of best practices and standards for industrial robotics. Groups like the Robotic Industries Association (RIA) and the International Federation of Robotics (IFR) provide valuable forums for industry stakeholders to collaborate on safety guidelines, technological standards, and ethical considerations. These organizations play a pivotal role in bridging the gap between regulatory frameworks and practical, real-world applications of robotics technologies.

Looking ahead, the regulatory environment for industrial robotics will continue to evolve in response to technological advancements and emerging use cases. As AI-powered humanoid robots become more prevalent in factories, the need for adaptive, forward-thinking regulations will be paramount. Regulatory bodies will need to collaborate closely with industry leaders, researchers, and policymakers to ensure that the standards in place are both rigorous and flexible enough to accommodate future innovations.

In conclusion, the current regulations in industrial robotics are foundational to ensuring the safe and effective integration of these technologies into manufacturing environments. While traditional safety standards lay the groundwork for mechanical and operational safety, emerging regulations are poised to address the unique challenges and opportunities presented by AI-driven systems. By fostering a robust regulatory framework, we can pave the way for a future where humanoid robots work alongside humans safely, efficiently, and harmoniously, revolutionizing the industrial landscape.

Developing Standards for AI Integration

As AI-powered humanoid robots continue to penetrate industrial sectors, developing robust and comprehensive standards for AI integration becomes essential. Standards ensure that these systems operate safely, effectively, and consistently across different environments. While the potential benefits of AI in industrial settings are vast, they can only be fully realized through well-defined regulatory frameworks that address various challenges and considerations.

The first step in developing these standards is understanding the multifaceted nature of AI integration. AI systems are not just software or hardware but a complex interplay of algorithms, data streams, sensor inputs, and real-time processing. This complexity necessitates a multi-layered approach to standard-setting, involving stakeholders from multiple disciplines, including computer science, engineering, ethics, and law.

One core area of focus is *interoperability*. For AI systems to be widely adopted, they must be able to communicate seamlessly with existing manufacturing systems and other robots. Standardized communication protocols and data formats are essential for this interoperability. This is where organizations like the International Organization for Standardization (ISO) can play a crucial role. They provide the necessary guidelines for establishing these protocols, ensuring that AI platforms from different vendors can work together effectively.

Another critical aspect of AI integration involves **safety standards**. AI-powered humanoid robots must be equipped with fail-safes and redundancy mechanisms to handle unexpected situations without causing harm. This involves not only hardware safety measures but also software-level checks and balances. Safety standards must be stringent, considering scenarios that traditional robots never encountered, such as decision-making in real time and operating in close proximity to human workers.

Ethical considerations also have to be part of the regulatory framework. Questions about accountability, transparency, and bias in AI systems need clear answers. Who is responsible if an AI system makes a faulty decision that leads to a loss or accident? How can we ensure that AI decisions are transparent and free of bias? These are complex questions but essential for public trust and acceptance. Developing ethical standards that outline guidelines for AI deployment and usage is indispensable.

Performance benchmarks for AI-powered robots will provide a yardstick against which innovations can be measured. Setting these benchmarks involves defining criteria for speed, accuracy, learning capability, and adaptability of these robots. Such benchmarks would facilitate fair competition among manufacturers while ensuring that the technology continues to advance.

Collaboration between governments, private sectors, and academia is pivotal in creating these standards. Governments can enact regulations, but they often lack the technical expertise to foresee every potential issue. Private sector companies have the expertise and practical experience but might prioritize profit over safety or ethical considerations. Academia can bridge this gap by providing research-based insights and serving as neutral ground for collaboration.

Global standards are another important aspect. In a connected world, where supply chains are international, having disparate standards can lead to inefficiencies and even failures. Establishing global standards ensures that AI-powered humanoids can be utilized seamlessly across different countries and industries, supporting international collaboration and trade. Organizations like ISO and the Institute of Electrical and Electronics Engineers (IEEE) have been instrumental in creating such global standards.

As we delve deeper into the specifics, *data security and privacy* emerge as vital components of AI standardization. AI systems rely heavily on data, making them susceptible to cyber threats. Security protocols must be established to protect sensitive information and ensure that AI systems remain resilient against potential cyberattacks. Integrating standards for data anonymization, encryption, and secure data transfer can mitigate these issues, safeguarding both the industry's and the workers' interests.

The **certification process** for AI systems in industrial automation has to be thorough and continuous. Once an AI system is certified, it doesn't mean it should not be reassessed. Given the rapid pace of AI advancements, standards should mandate periodic reviews and recertifications. This continuous evaluation ensures that the systems remain up-to-date and maintain the safety and efficiency expectations set out initially.

In the broader scope of things, public engagement is another crucial element. Educating the general public about the capabilities and limitations of AI can foster a culture of acceptance and understanding. This can be achieved through public consultations, workshops, and collaboration with educational institutions. Transparency in how AI systems are regulated, their success stories, and challenges can go a long way in building public trust.

Regulatory bodies should consider the concept of *sandbox environments* for AI testing. These controlled settings allow for real-world trials of AI-powered humanoids without the risk associated with a full-scale deployment. Insights gained from these sandbox tests can inform the development of more effective standards and guidelines, addressing issues that might not be apparent in theoretical or smaller-scale pilot tests.

Lastly, **adaptability of standards** is crucial. As the technology evolves, so too should the standards governing it. Regulatory frameworks must be flexible enough to accommodate new advancements and shifts in the industrial landscape. This calls for an ongoing dialogue between all stakeholders involved, ensuring that the standards evolve in tandem with technological progress.

In conclusion, developing standards for AI integration in industrial settings is a multifaceted process demanding a coordinated effort among various stakeholders. Interoperability, safety, ethics, global harmonization, security, and adaptability form the cornerstones of effective standardization. By focusing on these areas, we can pave the way for safe, efficient, and broadly accepted AI-powered humanoid robots in factories. This will not only enhance productivity but also shape a future where humans and robots can collaborate seamlessly, turning the vision of a technologically advanced industrial era into reality.

Chapter 13:
The Future of Humanoid Robotics in Factories

The horizon of humanoid robotics in factories is not just promising, but transformative. The integration of advanced AI and robotics is poised to reshape manufacturing by enhancing efficiency, reducing errors, and fostering a new era of human-robot collaboration. Factories will evolve into smart environments where humanoid robots make real-time decisions, adapt to new tasks, and seamlessly work alongside human workers. Emerging trends suggest that industries like electronics, pharmaceuticals, and logistics will witness significant adoption, driven by the need for precision and agility. As we look ahead, it becomes clear that the symbiosis of human intelligence and robotic precision will drive the next industrial revolution, optimizing production lines and setting new standards for performance and sustainability. This future will not merely be an extension of our current capabilities but a leap towards unprecedented innovation and efficiency in factory settings.

Predictions and Emerging Trends

The future of humanoid robotics in factories is a dynamic landscape marked by rapid advancements and transformative changes. As AI technology continues to evolve, it's becoming evident that humanoid robots will not just improve efficiency but will redefine traditional manufacturing paradigms. Several emerging trends point toward a future where humanoid robots are not just adjuncts but central players in industrial operations.

One of the most exciting predictions is the enhanced cognitive abilities of humanoid robots. Future iterations will likely exhibit advanced decision-making capabilities, powered by more sophisticated AI algorithms and neural networks. These robots will transition from mere task executors to intelligent decision-makers, capable of understanding complex scenarios and making real-time adjustments. This leap in AI cognition will significantly reduce downtime and increase productivity, as robots can troubleshoot and resolve issues autonomously.

Integration of machine learning and real-time data analytics into humanoid robots is another key trend. By leveraging vast amounts of data collected through sensors and IoT devices, robots will be able to predict system failures and maintenance needs before they become critical. Predictive analytics will empower factories to implement preemptive measures, thereby minimizing disruptions and ensuring smooth, uninterrupted production cycles. This capability is particularly valuable in industries with very high stakes on production continuity, such as automotive and electronics manufacturing.

Another interesting trend is the shift towards more intuitive human-robot interaction. Advances in natural language processing and emotional AI are expected to make these robots more user-friendly. Future humanoid robots will likely possess the ability to understand and respond to human emotions, making them effective collaborators in mixed human-robot teams. The notion of robots with emotional intelligence isn't just about recognizing and reacting to human emotions; it's about creating a symbiotic working relationship where robots can anticipate human needs and actions.

There's also a growing trend towards modularity and customization in the design of humanoid robots. Industries are demanding robots that can be easily reconfigured for different tasks and adapted to various working environments. Modular designs will allow for quick updates and changes, ensuring that robots can be tailored to specific industry needs. This adaptability could also extend to software updates, where new functionalities can be added without significant downtime or financial overhead.

Moreover, the future of humanoid robotics in factories will likely see increased collaboration between robots and humans. This isn't about replacing human labor but rather augmenting it. Robots will take over repetitive and physically demanding tasks, whereas humans will focus on roles that require critical thinking, creativity, and strategic decision-making. Such collaboration will not only improve efficiency but also enhance job satisfaction and workforce engagement.

Augmented reality (AR) and virtual reality (VR) technologies will also play a significant role in the interaction between humans and robots. By using AR and VR, human workers will be able to visualize robotic processes and provide input in a more intuitive manner. Training programs will also benefit from immersive AR and VR experiences, enabling workers to develop a deeper understanding of how to interact and collaborate with humanoid robots.

Robots as a Service (RaaS) could become a prevalent business model, providing more flexibility for companies looking to integrate humanoid robots. This model would allow businesses to lease robots on a subscription basis, making advanced robotics accessible to smaller companies that may not have the capital to invest in expensive robotic systems. The RaaS model facilitates regular upgrades and maintenance, ensuring that companies always have access to the latest technologies.

An important trend to watch is the role of 5G and edge computing in enhancing the capabilities of humanoid robots. With the higher data speeds and reduced latency offered by 5G, robots can communicate and make decisions in real-time with unprecedented efficiency. Edge computing will enable robots to process data locally, reducing the reliance on centralized cloud-based systems and thus minimizing delays. This combination of technologies will drive more responsive and efficient robotic systems, critical for time-sensitive manufacturing processes.

Environmental sustainability will continue to influence the development and deployment of humanoid robots. As industries increasingly focus on green manufacturing practices, robots will be designed to consume less energy and be made from recyclable materials. Moreover, robots will assist in the efficient use of resources, minimizing waste and optimizing the supply chain. Sustainable robotics will not only be a technological imperative but also a social responsibility.

In the manufacturing sector, we can expect to see a proliferation of humanoid robots across varied industries beyond automotive and electronics, expanding into sectors like pharmaceuticals, food processing, and logistics. Each industry will bring its own set of requirements and challenges, pushing the boundaries of robotic capabilities and applications. For instance, pharmaceutical manufacturing may require robots with precision in handling delicate substances, while the food industry might need robots that can safely manage perishable items under strict hygiene standards.

The regulatory landscape will evolve in tandem with humanoid robotics advancements. Governments and international bodies will likely establish new standards and certifications to ensure the safe and ethical deployment of these technologies in factories. These regulations will encompass areas such as robot safety, data protection, and ethical AI usage, ensuring a responsible and beneficial integration of humanoid robots into the industrial ecosystem.

Another fascinating prediction revolves around the concept of 'swarm robotics,' where multiple humanoid robots work in tandem, resembling a coordinated swarm. This approach can greatly enhance efficiency and adaptability in complex production environments. Swarm robotics will enable factories to scale operations easily, with robots dynamically adjusting to new tasks and collaborating seamlessly to achieve collective goals.

Advanced humanoid robots will also drive innovations in supply chain management. Equipped with AI and blockchain technologies, these robots will bring transparency and traceability to the supply chain, ensuring that every step, from raw material procurement to final product delivery, is efficiently managed. This kind of operational clarity is invaluable for quality assurance and regulatory compliance.

In conclusion, the future of humanoid robotics in factories is not just a tale of technological advancement but a narrative of transformation. With predictions pointing towards robots with enhanced cognitive abilities, intuitive human-robot interaction, modular designs, and significant roles in sustainability and supply chain management, the industrial landscape is set to undergo a revolution. As these trends unfold, the amalgamation of human creativity and robotic precision will unlock new realms of efficiency and innovation, making the future both exciting and promising.

Potential Industries for Expansion

The possibilities for expansion of humanoid robotics into new industries are vast and varied. While the automotive and manufacturing sectors have traditionally been at the forefront of robotic integration, the unique capabilities of humanoid robots—enhanced dexterity, improved decision-making through AI, and the ability to operate in human-centric environments—make them highly suitable for numerous other fields.

One promising area is the logistics and warehousing industry. Given the continual rise in e-commerce, warehouses must manage larger volumes of goods with increasing efficiency and speed. Humanoid robots can perform a range of tasks, from item picking and packing to inventory management and quality control. Their adaptability allows them to be redeployed as needs fluctuate, an essential feature in dynamic warehouse environments.

Healthcare is another industry ripe for the introduction of humanoid robots. Humanoids can assist with numerous tasks, such as transporting medical supplies within a hospital, providing patient monitoring, and even conducting basic diagnostic procedures. In surgical settings, humanoid robots equipped with advanced AI can assist surgeons by providing steadier, more precise control over instruments.

Humanoid robots could significantly impact the construction industry as well. Their ability to handle complex tasks and navigate irregular terrains makes them excellent candidates for roles that require high precision and labor-intensive work. They could be deployed for tasks that are too dangerous or monotonous for human workers, such as high-altitude installations or repetitive bricklaying. Innovations in AI have also enabled robots to learn from human architects and engineers, ensuring that they can adhere to design specifications accurately.

Retail is another sector where humanoid robots could bring about transformative changes. Retail environments often require personalized customer service, inventory control, and merchandising—tasks that humanoid robots are increasingly capable of performing. They can greet customers, offer product recommendations based on AI-driven data analytics, and even assist in checkouts. In a world where customer experience is paramount, humanoid robots can provide a level of assistance and interaction that is consistent and engaging.

In the agriculture sector, humanoid robots offer a new frontier for innovation. Farming involves a variety of tasks that can be physically demanding and time-consuming, such as planting, harvesting, and sorting crops. Humanoids, equipped with machine learning algorithms, can analyze soil conditions, monitor crop health, and apply precision agriculture techniques to optimize yields.

Another intriguing field is the hospitality industry. Hotels and restaurants are starting to experiment with humanoid robots to handle reception tasks, room service, and even food preparation. These robots can provide consistent service, minimize human error, and work around the clock, enhancing overall operational efficiency. Guests could enjoy faster check-ins and personalized recommendations for their stay, creating a more tailored and memorable experience.

The education sector also stands to benefit considerably from humanoid robots. In classrooms, robots can assist teachers by handling administrative tasks, providing tutoring in subjects where students need additional help, and creating an engaging learning environment through interactive lessons. This support allows teachers to focus on one-on-one interactions and curricular development, ultimately enriching the educational experience.

In public safety and emergency services, humanoid robots can play crucial roles. They can be deployed in disaster zones where human access is limited or too dangerous. Tasks like searching for survivors, transporting supplies, and communicating with trapped individuals can be managed more effectively by humanoid robots equipped with advanced sensors and AI for real-time decision-making.

Exploring the possibilities in entertainment, humanoid robots can be utilized for performing arts, acting in movies, or even as companions in theme parks. Their ability to mimic human gestures and expressions can bring new dimensions to storytelling and interactive experiences, captivating audiences in unprecedented ways.

Moreover, spaces like offices and high-tech industries could benefit from humanoid robots. From managing administrative tasks to acting as intelligent personal assistants, humanoid robots could change the way businesses operate. With capabilities such as real-time data analysis and problem-solving, they can offer invaluable support, especially in high-stakes decision-making processes.

In summary, the potential for humanoid robots to expand into various industries is not just feasible but highly promising. As AI and robotics technologies continue to evolve, the versatility and adaptability of these robots will enable them to integrate seamlessly into new environments, fundamentally altering the way work is performed across multiple sectors. While there will be challenges, such as ensuring safety standards and addressing ethical concerns, the opportunities for improved efficiency, lower operational costs, and enhanced human experiences are immense. The future of humanoid robotics in diverse industries is not just a possibility—it's an imminent reality.

Chapter 14:
Designing Robots for Enhanced Collaboration

In the age of AI-powered humanoid robots, designing for enhanced collaboration isn't just about technical excellence—it's about creating machines that humans can work with intuitively and efficiently. To achieve this, ergonomics and human factors engineering play crucial roles, ensuring that robots are built to complement human physical capabilities and limitations. Moreover, interactive interfaces and user experience design are key to seamless interactions, allowing for more natural and effective communication between humans and robots. These elements empower workers to leverage the full potential of humanoids, transforming factory floors into spaces where man and machine work in concert, effortlessly pushing the boundaries of productivity and innovation. By integrating advanced technologies with a profound understanding of human needs, we can create robots that not only enhance industrial operations but also foster a more collaborative and adaptive workplace.

Ergonomics and Human Factors Engineering

In the domain of designing robots for enhanced collaboration, ergonomics and human factors engineering play a pivotal role. These fields focus on optimizing the interaction between humans and robots to ensure that both parties can perform their functions safely, efficiently, and comfortably. Employing a human-centered design approach, engineers and designers work together to understand the physical, cognitive, and organizational aspects that influence how humans interact with robotic systems in the industrial setting.

A key factor in this process is the ergonomic design of humanoid robots, which are intended to operate in environments originally designed for human workers. By mimicking human anatomy, humanoid robots can more easily adapt to these pre-existing workspaces. This not only reduces the time and cost needed for reconfiguring manufacturing lines but also minimizes disruptions to current workflows. Integrating robots that can seamlessly step into these roles ensures smoother adoption and more immediate benefits for industries.

Drawing from principles of anthropometry, designers must consider the various physical dimensions and capabilities of human operators. Elements such as reach, grip strength, and visual field are essential when creating robots that can assist or replace human workers. For instance, a robot's arm must be able to move as fluidly and precisely as a human arm, while also having the strength to perform tasks that may require significant force. Additionally, these robots need to be adaptable to different individuals within a workforce, necessitating adjustable features or a degree of customization.

On the cognitive side, successful robotics design hinges on understanding how humans process information and make decisions. Human factors engineering addresses these considerations by ensuring that robots can communicate effectively with human workers. This might involve the use of intuitive user interfaces, clear visual and auditory signals, and even predictive analytics to anticipate and react to human actions. Such features enhance the collaborative experience, making it more natural and efficient.

One significant challenge in human-robot interaction is managing the cognitive load on human operators. When robots and humans collaborate, the user interfaces must be designed to be intuitive, minimizing the learning curve and reducing the risk of errors. To achieve this, designers often employ iterative design processes that include user feedback at various stages. By involving actual end-users in the design process, engineers can create systems that workers find more intuitive and easier to use.

Safety is another critical aspect of ergonomics and human factors engineering in robotic design. Robots must be designed to operate safely around humans, which involves incorporating advanced sensor technologies capable of detecting and responding to human presence. Proximity sensors, force feedback mechanisms, and vision systems all contribute to a robot's ability to operate without posing a risk to nearby human workers. Engineers must account for worst-case scenarios to ensure that even in unexpected situations, both humans and robots remain safe.

Designing user interfaces also falls under the umbrella of human factors engineering. When humans interact with robots, whether directly or via control panels, the interfaces must be clear, usable, and responsive. This could mean designing graphical user interfaces (GUIs) that are easy to understand or developing voice interaction systems that allow for hands-free control. The end goal is always to facilitate smoother and more effective interactions, enabling humans to harness the full potential of their robotic colleagues.

Furthermore, the emotional and psychological impacts of working alongside robots must be considered. Human factors engineering can aid in mitigating any feelings of unease or discomfort that might arise from robotic integration. By designing robots that are approachable, with physical features and movements that are less mechanical and more human-like, engineers can foster a more positive work environment. This is particularly important in maintaining high morale and productivity levels among human workers.

In collaborative settings, the synergy between humans and robots is paramount. Training programs play a crucial role in preparing human workers to interact effectively with their robotic counterparts. These programs are often designed incorporating principles from both ergonomics and human factors engineering, focusing on physical ease and cognitive clarity. Ensuring that employees are comfortable and confident in their interactions with robots leads to better cooperation and productivity.

Monitoring and continual assessment are also parts of the ergonomic and human factors approach. Regular feedback loops, use of performance metrics, and direct observations help in making ongoing adjustments to improve interactions. This iterative process supports the continuous optimization of both robotic systems and human workflows, increasing overall efficiency and safety.

Lastly, addressing ergonomics and human factors is not a one-time action but an ongoing commitment. As technologies evolve and new opportunities for human-robot collaboration arise, continuous research and development are necessary. Staying ahead of ergonomic needs and human factors considerations ensures that industries can adapt more fluidly to changes, maintaining a high standard in worker well-being and operational efficiency.

In conclusion, ergonomics and human factors engineering provide a foundational framework for designing effective and harmonious human-robot collaborations. By prioritizing the comfort, safety, and intuitive interaction of human workers, engineers can create robots that truly enhance the industrial environment. This holistic approach not only optimizes productivity but also ensures a sustainable and positive integration of AI-powered humanoid robots into the workforce.

Interactive Interfaces and User Experience

In the evolving landscape of industrial automation, humanoid robots are increasingly taking center stage. The sophistication of these robots relies heavily on interactive interfaces and their ability to offer a seamless user experience. A well-designed interface, coupled with user-friendly experiences, isn't just a luxury but a necessity for fostering collaboration between humans and machines. When people and robots work together, the interface essentially serves as a bridge that connects human intuition with robotic precision, enhancing efficiency and fostering symbiotic growth.

Humanoid robots, equipped with interactive interfaces, are designed to respond to human inputs more intuitively and predictably. This aspect can be a game-changer on factory floors where quick adaptation is often the key to maintaining production efficiency. For instance, touchscreens with intuitive layouts, voice-activated commands, and gesture recognition systems are some interfaces helping to make interactions smoother and more natural. Imagine an operator gesturing to a robot to pick up a part or issuing a voice command to start a diagnostic process—these seemingly simple actions can significantly reduce operational delays and enhance productivity.

The user experience, however, goes beyond just the interface. It encompasses the entire interaction lifecycle between humans and robots. An interactive interface must be ergonomic, ensuring that it minimizes physical strain and cognitive load on human workers. It should also be adaptable, learning from user behavior and improving over time. For example, a robot that can adjust its speed and movement based on the familiarity and proficiency of the human worker handling it offers a more personalized and efficient user experience.

One of the crucial aspects of successful interactive interfaces is their capacity for real-time feedback. Factory environments are typically dynamic, requiring adjustments on the fly. Therefore, interfaces that provide instant feedback ensure that operations run smoothly. Visual indicators, auditory signals, and haptic feedback can all contribute to this real-time communication. A touch screen that changes color to indicate an error, or a beep sound from the robot signaling task completion can help in maintaining the workflow without disruptions.

Another significant facet is the emotional and psychological comfort of human workers as they adapt to collaborating with robots. Establishing trust is essential for effective human-robot interaction. Interactive interfaces can aid in this by being transparent about the robot's functions and intentions. Progress indicators, status updates, and clear error messages help in making the interactions predictable and thereby trustworthy. When workers feel confident about what the robot is doing, and why it is doing it, they are more likely to accept it as a reliable team member rather than a threatening entity.

Furthermore, advances in AI have made conversational interfaces more potent and practical. Powered by advanced Natural Language Processing (NLP) algorithms, these interfaces allow robots to understand and execute complex verbal instructions. Imagine issuing a series of commands to a humanoid robot in plain English and seeing them executed accurately—this not only amplifies efficiency but also reduces the learning curve for workers. The less time workers spend learning how to communicate with robots, the more time they can devote to complex tasks that require human ingenuity.

However, the design of effective interactive interfaces also requires a deep understanding of the job roles and tasks in specific industries. Customizing interfaces to fit the needs of different sectors ensures that robots can be more effectively integrated into diverse workflows. For instance, the interface requirements for a robot working in an automotive assembly line will significantly differ from those in a pharmaceutical manufacturing setup. Tailoring these interfaces allows for smoother operations and higher productivity by accommodating industry-specific nuances.

But interactive interfaces aren't just about touchscreens or voice commands. The integration of Augmented Reality (AR) and Virtual Reality (VR) can transform how humans and robots interact. AR glasses that display real-time data feeds from robots, or VR headsets that allow for immersive training sessions, provide layers of interaction that were previously unimaginable. These technologies enable workers to engage with robots in a more informed and comfortable manner, thereby reducing errors and enhancing overall operational efficiency.

Simplifying the user experience can also make a massive difference in error reduction and safety, both crucial in an industrial environment. An interface that is cluttered or overly complicated can lead to mistakes, jeopardizing both human and robot safety. Clear, straightforward designs that include essential features such as emergency stops, easy access to manuals, and simple diagnostic tools ensure that even those less familiar with advanced technology can operate robots effectively and safely.

Additionally, interactive interfaces can promote inclusivity, making it easier for a diverse workforce to adapt to new technologies. Features like multi-language support, variable text sizes, and customizable user profiles can make these interfaces accessible to workers from different backgrounds and with varying levels of technical expertise. This inclusivity ensures that the benefits of humanoid robots are accessible to all, not just a select few.

As we look towards the future, the convergence of AI with IoT (Internet of Things) can further elevate interactive interfaces into hyper-intelligent systems capable of predictive analytics and autonomous decision-making. By continuously gathering and analyzing data in real-time, these advanced interfaces can proactively assist operators, foresee potential issues, and suggest corrective actions before any disruption occurs. This not only enhances operational efficiency but also contributes to a more proactive and resilient production environment.

Ultimately, the goal of designing interactive interfaces and enriching user experiences is to create a harmonious ecosystem where humans and robots function as cohesive units. This synergy is vital not just for operational efficiency, but also for ensuring that human workers feel empowered rather than displaced by their robotic counterparts. In doing so, industries can harness the full potential of humanoid robots, paving the way for a future where technology complements human effort rather than competing with it.

Interactive interfaces and user experiences will continue to evolve, driven by technological advancements and user feedback. The successful implementation of these interfaces will be a critical determinant in how effectively humanoid robots can be integrated into industrial settings. The future of industrial automation lies not just in the capabilities of the robots themselves but in the ease and intuitiveness with which humans can interact with these machines. As we move forward, the ongoing refinement of these interactive interfaces will play a pivotal role in shaping the next era of collaborative manufacturing.

Chapter 15:
The Role of Big Data and Analytics

Big Data and Analytics are reshaping the industrial landscape by enabling the full potential of AI-powered humanoid robots. With vast amounts of data generated in manufacturing environments, companies can now leverage analytics to unlock unprecedented insights into their operations. This data-driven approach allows for real-time monitoring, predictive maintenance, and advanced troubleshooting, significantly enhancing efficiency and productivity. By analyzing patterns and anomalies, manufacturers can anticipate issues before they occur, reducing downtime and mitigating risks. The integration of Big Data with AI not only optimizes robot performance but also supports continuous improvement processes, driving innovation and maintaining a competitive edge. As factories become smarter and more interconnected, the role of data analytics will continue to be a cornerstone in the evolution of industrial automation, ensuring that humanoid robots can adapt and thrive in dynamic production environments.

Leveraging Data for Improved Performance

In the ever-evolving landscape of industrial automation, the role of big data and analytics cannot be overstated. "Leveraging Data for Improved Performance" dives into how AI-powered humanoid robots utilize vast amounts of data to optimize their functionalities, ultimately revolutionizing manufacturing processes. This section illuminates the interplay between raw data, sophisticated algorithms, and actionable insights, showcasing the transformative power of data in the realm of industrial robotics.

Big data is characterized by its volume, velocity, and variety. Humanoid robots in factories generate and consume a staggering amount of data from a myriad of sources, including sensors, cameras, and other embedded devices. This data forms the backbone of their learning and decision-making processes, allowing them to adapt and improve continuously. In a sense, these robots are data-driven entities, constantly ingesting, processing, and analyzing information to refine their actions.

One of the primary ways data is leveraged is through predictive analytics. Predictive models, built from historical data, enable robots to foresee potential issues and take preemptive actions. For instance, if data indicates that a particular machine component is likely to fail soon, the humanoid can either alert human supervisors or initiate maintenance protocols autonomously. This capability reduces downtime and enhances overall productivity.

Moreover, real-time data processing is crucial for maintaining the efficacy of humanoid robots. Manufacturing environments are dynamic, and decisions often need to be made in milliseconds. Advanced data analytics platforms can process incoming data streams on the fly, enabling robots to make informed decisions almost instantaneously. This means better-quality products, faster production times, and more efficient use of resources.

The implementation of machine learning algorithms allows humanoid robots to learn from data continuously. Through observational learning and pattern recognition, these robots can fine-tune their actions based on both past experiences and real-time data. Over time, this iterative process results in improved performance, greater accuracy, and reduced error rates. The ability to learn and adapt makes humanoid robots invaluable assets in the pursuit of operational excellence.

Data quality is another critical aspect. The efficacy of any analytical model is directly proportional to the quality of the data it utilizes. Therefore, robust data governance frameworks are essential. Such frameworks ensure the accuracy, integrity, and relevance of data, laying the foundation for reliable analytics. Clean, high-quality data enhances the robots' ability to make precise decisions, elevating the entire manufacturing process.

Furthermore, the integration of data from various sources allows for a more holistic view of factory operations. By correlating data from different machines, processes, and stages of production, you gain deeper insights into bottlenecks, inefficiencies, and opportunities for improvement. This multi-source data aggregation paints a comprehensive picture, empowering robots and human supervisors alike to make well-informed decisions.

Big data analytics also plays a pivotal role in optimizing supply chain logistics. By analyzing data related to material availability, lead times, and demand patterns, humanoid robots can make real-time adjustments to production schedules. This optimization reduces transit times, cuts costs, and ensures timely delivery of products. In a globalized economy where supply chain efficiency is paramount, such capabilities offer a competitive edge.

One cannot overlook the privacy and ethical considerations inherent in data utilization. With vast amounts of data being collected, it's imperative to address issues such as data security and user consent. Employing robust encryption methods and complying with regulatory frameworks ensures that data is handled responsibly, mitigating risks and building trust in AI systems.

Another fascinating application of data is in the realm of quality control. Traditional quality assurance methods often involve manual inspections, which can be time-consuming and prone to human error. Humanoid robots, equipped with advanced data analytics tools, can perform these inspections with unparalleled precision. By analyzing data from sensors and cameras, they can identify defects and deviations that might be invisible to the human eye. This leads to higher-quality products and reduced waste.

Additionally, data analytics facilitates the continuous improvement of manufacturing processes. By analyzing operational data, factories can identify trends, measure performance, and implement changes that drive efficiency. Continuous monitoring and feedback loops create a culture of perpetual enhancement, aligning with the principles of lean manufacturing and Six Sigma methodologies.

Humanoid robots also excel in task optimization through data analysis. By evaluating the data on how various tasks are performed, they can suggest more efficient methods or even reconfigure their own workflows. For example, if data reveals that a specific assembly task can be performed faster by altering the sequence of actions, the robot can adapt accordingly. This flexibility is crucial for maintaining high levels of performance in a constantly changing industrial environment.

Furthermore, leveraging big data paves the way for innovation. The insights gleaned from data analytics can inspire new manufacturing techniques, product designs, and business strategies. In essence, data becomes a catalyst for creativity and innovation, driving the industry forward. Companies that harness the full potential of their data assets will likely stay ahead of the curve, leading the charge in industrial transformation.

Collaboration between humans and robots also benefits immensely from data analytics. By sharing data in real-time, both parties can synchronize their efforts more effectively. Human workers can receive data-driven insights that enhance their decision-making, while robots can adjust their actions based on real-time feedback from humans. This synergy improves overall productivity and fosters a more harmonious working environment.

Finally, the scalability of data analytics solutions means that they can be customized to fit factories of all sizes. From small workshops to large-scale manufacturing plants, data-driven strategies can be tailored to meet specific needs and objectives. This scalability ensures that the benefits of leveraging data for improved performance are accessible to a broad spectrum of businesses, democratizing the advantages of advanced analytics.

In summary, the role of big data and analytics in enhancing the performance of AI-powered humanoid robots is transformative. By harnessing the power of data, these robots can optimize workflows, predict and prevent issues, improve quality, and inspire innovation. As the industrial landscape becomes increasingly data-driven, the ability to effectively leverage this data will determine the future success of manufacturing operations, paving the way for a new era of efficiency and productivity.

Predictive Maintenance and Troubleshooting

In the evolving landscape of industrial automation, predictive maintenance and troubleshooting have become pivotal elements under the umbrella of big data and analytics. With AI-powered humanoid robots intricately woven into the fabric of modern factories, the need for efficient maintenance strategies becomes more critical. Predictive maintenance, as distinguished from reactive or scheduled maintenance, leverages data analytics and machine learning algorithms to forecast when equipment failures might occur. This proactive approach minimizes downtime and maintenance costs, significantly enhancing productivity and efficiency.

Traditional maintenance systems relied heavily on routine inspections and reactive repairs. These models often resulted in unplanned downtimes and inefficient resource allocation. With big data streaming in from a myriad of sensors embedded in humanoid robots, predictive maintenance is not merely a possibility but a necessity. AI algorithms analyze vast datasets to identify patterns, anomalies, and potential points of failure, creating a more responsive and resilient manufacturing environment.

The implementation of predictive maintenance strategies in robotic systems involves several steps. First, sensors collect real-time data on various machine parameters, including temperature, vibration, and electrical signals. This data is then integrated and processed via sophisticated machine learning models. These models, trained to detect deviations from the norm, trigger alerts before a breakdown occurs. Consequently, technicians can perform maintenance activities exactly when necessary, avoiding the costly repercussions of unexpected failures.

For example, in an automotive plant, humanoid robots may exhibit early signs of wear in their actuators through subtle changes in vibration patterns. Predictive analytics can detect these variations and forecast potential failure weeks before it happens, allowing the replacement of parts during scheduled downtimes. This ensures that the production line remains uninterrupted and the workforce can plan their schedules effectively.

Moreover, predictive maintenance enhances the overall lifespan of factory equipment. By attending to potential issues promptly, businesses can avoid the severe damage that unchecked faults might cause. Over time, this results in a substantial cost saving, as equipment doesn't need to be replaced as frequently. The cumulative benefits of predictive maintenance translate into a more sustainable and economically viable manufacturing operation.

On the troubleshooting front, AI-powered humanoid robots are redefining how factories address malfunctions. Traditionally, identifying and fixing issues required extensive human intervention and, often, the presence of specialized technicians. Today, robots equipped with advanced diagnostics capabilities can autonomously detect and even resolve certain malfunctions. By conducting self-assessments and identifying the root causes of anomalies, these robots streamline the troubleshooting process.

The integration of natural language processing (NLP) and machine learning models in humanoid robots allows for more sophisticated interaction with human operators. When a problem arises, these robots can explain the issue, propose potential solutions, or guide technicians through the troubleshooting steps. This interaction reduces the complexity of repairs and ensures that even less experienced staff can effectively manage the equipment.

Additionally, centralized data analytics platforms are becoming integral to modern manufacturing ecosystems. These platforms aggregate data from multiple robots and other factory equipment, providing a holistic view of operational health. When an issue is detected in one robot, similar data patterns can be analyzed across the factory floor to prevent the spread of the problem. This creates a more resilient and robust maintenance strategy.

One noteworthy advantage of this approach is the ability to harness historical data for learning and improvement. By continuously analyzing past incidents, machine learning models refine their predictive accuracy and evolve their troubleshooting protocols. This iterative process enhances the reliability of humanoid robots, reducing the frequency and severity of breakdowns over time.

Predictive maintenance and troubleshooting, when combined with big data analytics, offer a significant competitive edge for industries adopting AI-powered humanoid robots. Organizations can capitalize on these technologies to optimize operations and reduce costs, all while maintaining high levels of efficiency and reliability. Furthermore, the reduced downtime and seamless troubleshooting processes contribute to a more productive and streamlined manufacturing environment.

In conclusion, the integration of big data and analytics in predictive maintenance and troubleshooting is a transformative development in industrial automation. By proactively addressing potential failures and leveraging AI for advanced diagnostics, factories can achieve unmatched levels of efficiency and reliability. As the technology continues to evolve, its applications will undoubtedly expand, further cementing the role of AI-powered humanoid robots in the future of manufacturing.

Chapter 16:
Training and Skill Development for the Future Workforce

In an era where AI-powered humanoid robots are redefining industrial landscapes, equipping the workforce with the right skills has never been more crucial. As automation takes on more complex tasks, workers must shift from manual operations to roles that require advanced cognitive abilities and technical acumen. Companies are investing heavily in educational programs and industrial training initiatives that focus on robotics and AI, fostering a culture of continuous learning and adaptability. These programs are not just about understanding the machinery but learning to collaborate effectively with intelligent systems. By bridging the gap between human expertise and robotic efficiency, we can create an environment where technological advancements drive productivity while empowering employees with new opportunities. The future workforce will not only need to be proficient in the latest technologies but also possess the capability to innovate and thrive in an ever-evolving industrial ecosystem.

Educational Programs for Robotics and AI

Training and skill development for the future workforce must start early and be comprehensive to produce individuals proficient in robotics and AI. Educational programs need to span different educational levels and institutions, tailored to address the complexities and rapid advancements within the field. Starting from elementary schools to higher education and professional training centers, the initiatives for robotics and AI education must be diversified and extensive.

Elementary education forms the bedrock for nurturing interest in STEM fields, where foundational concepts of robotics and AI can be introduced through interactive and playful methods. Schools across various regions have started integrating robotics kits and basic programming languages such as Scratch into their curriculums. Competitions like First LEGO League provide children with platforms to explore, experiment, and showcase their creations, fostering a culture of curiosity and problem-solving from a young age.

As students progress to middle and high school, the exposure needs to intensify with more structured coursework and hands-on projects. These programs should introduce them to the fundamentals of AI, machine learning, and more sophisticated robotic systems. Schools can collaborate with tech companies and universities to offer workshops, summer camps, and mentorship programs, allowing students to gain insights from industry professionals and academics. The goal at this stage should be to deepen their understanding and build solid technical skills that can serve as a springboard for advanced education and careers in robotics and AI.

Higher education institutions play a pivotal role by offering specialized degrees and research opportunities in robotics and AI. Universities must develop interdisciplinary programs that blend electrical engineering, computer science, mechanical engineering, and cognitive sciences to produce well-rounded professionals. Labs dedicated to robotics experimentation provide students with the environment and tools to innovate and refine their ideas. Universities like Carnegie Mellon, MIT, and Stanford have leading-edge programs and facilities focused on producing breakthroughs in robotics and AI, setting benchmarks for educational excellence worldwide.

Vocational training and community colleges also have significant roles to play. Not everyone needs to pursue a four-year degree to contribute effectively to the robotics and automation sectors. Programs geared toward specific skills, such as robotic system maintenance, AI-powered machinery operation, and programming, can be highly beneficial. These shorter, intensive courses are designed to meet the immediate needs of the industry, ensuring that the workforce remains adaptable and ready to implement and manage new technologies as they emerge.

Online education platforms like Coursera, edX, and Udacity have democratized access to high-quality education in robotics and AI. These platforms offer courses designed by top universities and industry leaders, making it possible for anyone with an internet connection to learn and upgrade their skills. Bootcamps and nano-degree programs focus on practical knowledge and real-world applications, preparing learners for the challenges they will face in their careers. A blend of asynchronous learning, live sessions, and project-based assessments ensures a comprehensive understanding and hands-on experience.

Corporate training programs are also vital in ensuring the current workforce can transition and keep pace with technological advancements. Many companies now offer in-house training modules focusing on upskilling employees in robotics and AI. Additionally, industry partnerships with educational institutions help tailor coursework to meet specific organizational needs, ensuring that what is taught is directly relevant to the work environment employees will encounter.

Certification programs offered by professional bodies, such as the Institute of Electrical and Electronics Engineers (IEEE) or the

Robotics Industries Association (RIA), lend significant credibility. Recognized and respected in the industry, these certifications validate an individual's proficiency and readiness to tackle complex problems in robotics and AI. They also serve as motivators for continuous learning and professional development.

Furthermore, industry competitions and hackathons have become instrumental in driving innovation and skill development. These events provide a unique platform where students, hobbyists, and professionals converge to solve real-world problems using robotics and AI technologies. Participants often work under time constraints and with limited resources, which fosters creativity, teamwork, and rapid problem-solving skills. Many groundbreaking ideas and prototypes have emerged from these events, underscoring their importance in the educational ecosystem.

International collaborations also enhance educational programs significantly. Initiatives like exchange programs, joint degrees, and collaborative research projects between institutions from different countries expose students to diverse perspectives and methodologies. This global approach prepares students to work in an international market, understanding various standards and practices in robotics and AI.

Scholarships and funding from both government and private sectors play an essential role in making education in robotics and AI accessible to a broader demographic. Grants and financial aid make it possible for students from underrepresented or economically disadvantaged backgrounds to pursue education without the burden of financial constraints. It is crucial to ensure that diversity in these fields mirrors the diversity of society at large, fostering innovation from various life experiences and viewpoints.

Lastly, continuous curriculum updates are imperative. Given the fast-paced evolution of AI and robotics technologies, educational institutions must regularly review and update their coursework to stay relevant. This requires close communication with industry leaders to anticipate trends and skills in demand, ensuring that students are always on the cutting edge of technological advancements.

By investing in these various educational programs and pathways, we can build a robust pipeline of talent well-versed in robotics and AI. This not only addresses the current skills gap but also propels innovation, ensuring that the workforce is equipped to tackle the challenges and harness the opportunities of the future industrial landscape. Training and skill development in robotics and AI is not just a preparation for tomorrow's jobs; it's shaping the very fabric of our technological future.

Industrial Training Initiatives

Industrial training initiatives are indispensable in preparing the future workforce for the evolving landscape of manufacturing. As humanoid robots increasingly integrate into production lines, the need for specialized training programs becomes more evident. These initiatives focus on equipping workers with the skills necessary to collaborate effectively with advanced AI-powered systems, ultimately leading to enhanced productivity and innovation within industrial settings.

One of the primary objectives of these training programs is to bridge the knowledge gap between traditional manufacturing methods and the new, technology-driven approaches. This involves a comprehensive curriculum that covers various aspects of robotics, AI, and machine learning. By understanding the core functionalities and operations of humanoid robots, employees can better adapt to their roles in modern factories. Industrial training initiatives often incorporate both theoretical learning and hands-on practical experience, thereby ensuring a well-rounded education.

To address the diverse needs of the workforce, these programs are designed to be flexible and modular. For instance, training sessions can be customized to suit the specific requirements of different industries, such as automotive, electronics, or pharmaceuticals. This adaptability allows organizations to tailor their training efforts to the unique challenges and opportunities presented by their respective sectors.

An essential component of these initiatives is the emphasis on continuous learning and development. Given the rapid advancements in AI and robotics, staying updated with the latest technologies is crucial. Industrial training programs often include ongoing workshops, seminars, and refresher courses to keep employees abreast of new developments. This commitment to lifelong learning not only benefits individual workers but also contributes to the overall competitiveness and resilience of the industry.

Another critical aspect of industrial training initiatives is fostering collaboration between humans and robots. Effective teamwork between human workers and AI-powered machines can significantly enhance operational efficiency. Training programs focus on developing soft skills like communication, problem-solving, and teamwork, which are vital for successful human-robot collaboration. By cultivating a culture of mutual respect and understanding, these initiatives aim to create a harmonious and productive work environment.

Moreover, industrial training initiatives often partner with educational institutions, research centers, and technology providers to offer comprehensive training solutions. Such collaborations ensure that the training content is aligned with cutting-edge research and industry best practices. By leveraging the expertise of various stakeholders, these programs can deliver high-quality education that meets the evolving needs of the workforce.

When it comes to the format of these training initiatives, a blend of traditional classroom instruction and digital learning platforms is commonly employed. E-learning modules, virtual simulations, and interactive tutorials provide a flexible and accessible way for employees to acquire new skills. Additionally, the use of advanced technologies like augmented reality (AR) and virtual reality (VR) in training programs allows for immersive and interactive learning experiences, which can be particularly effective in complex or hazardous work environments.

Industrial training initiatives also play a crucial role in promoting innovation within organizations. By encouraging a mindset of experimentation and continuous improvement, these programs help workers develop the confidence and skills to explore new solutions and optimize existing processes. This culture of innovation is essential for maintaining a competitive edge in the fast-paced world of manufacturing.

Furthermore, these training programs are not limited to technical skills alone. They also address the broader implications of integrating humanoid robots into the workforce, such as ethical considerations, workplace safety, and regulatory compliance. By providing a holistic education, industrial training initiatives ensure that employees are well-prepared to navigate the complexities of modern manufacturing environments.

To measure the effectiveness of industrial training initiatives, companies often implement robust evaluation and feedback mechanisms. By assessing the impact of training programs on employee performance and productivity, organizations can identify areas for improvement and make data-driven decisions. Regular feedback from participants also helps in refining the training content and delivery methods, ensuring that the programs remain relevant and effective.

In conclusion, industrial training initiatives are vital in equipping the future workforce with the skills and knowledge required to thrive in a technologically advanced manufacturing landscape. By focusing on continuous learning, collaboration, and innovation, these programs not only enhance individual capabilities but also drive the overall growth and competitiveness of the industry. As AI-powered humanoid robots continue to revolutionize industrial production, the importance of comprehensive and adaptive training programs cannot be overstated.

Chapter 17:
Robotics in a Globalized Economy

As the world becomes increasingly interconnected, the role of robotics in the globalized economy is more crucial than ever. International collaborations are sprouting up, with countries sharing expertise to set universal standards that ensure both innovation and safety. Case studies from diverse regions show how AI-powered humanoid robots are being tailored to meet local business needs while adhering to global best practices. This exchange paves the way for a more synchronized and efficient global manufacturing landscape. By leveraging the collective intelligence of different nations, industries can optimize production, reduce costs, and foster a culture of continuous improvement, all while maintaining competitiveness on the world stage. Robotics, therefore, isn't just a technological evolution; it's a transformative force that's bringing the world closer through shared progress and collaborative innovation.

International Collaborations and Standards

In today's interconnected world, the momentum of technological advancement is undoubtedly propelled by international collaborations, especially in the realm of robotics. As industrial automation climbs new heights, it's evident that AI-powered humanoid robots are playing a pivotal role in shaping the global workforce. Countries and companies from various corners of the globe are collaborating to set the tone for the future of robotics, ensuring that advancements are not just innovative but also standardized for universal adoption.

International collaboration in robotics enables the pooling of resources and knowledge from diverse regions. It's more than just sharing technical expertise; it involves harmonizing different cultural perspectives on technology and work. Together, these partnerships create robust frameworks that balance innovation with ethics, safety, and efficiency.

Establishing international standards for humanoid robots is crucial to fostering a uniform level of safety, functionality, and interoperability. Standards bodies like the International Organization for Standardization (ISO) and the Institute of Electrical and Electronics Engineers (IEEE) play a significant role. By developing comprehensive guidelines and specifications, these organizations ensure that robotic systems from different manufacturers can work seamlessly together, regardless of their country of origin. These standards cover everything from hardware and software to protocols for communication and safety protocols.

One of the major benefits of international standards is that they help mitigate the risks associated with the deployment of AI-powered humanoid robots. By adhering to globally accepted safety norms, companies can significantly reduce the risk of malfunctions and accidents. This not only secures worker safety but also protects the company's investment in advanced automation technology.

Additionally, international standards foster a competitive yet cooperative environment. Countries and corporations can freely innovate within the constraints of these standards, ensuring that new developments are backward-compatible and forward-thinking. This collaborative spirit accelerates technological breakthroughs, making advanced robotics accessible to a broader market.

Consider, for instance, the collaborative efforts between Japan, a leader in robotics, and Germany, an industrial powerhouse. These nations have spearheaded multiple joint ventures to create humanoid robots that blend precision engineering with advanced AI. Their concerted efforts have led to the development of robots capable of performing complex tasks with unmatched accuracy and reliability. For instance, Japan's focus on meticulous craftsmanship complements Germany's expertise in efficient, large-scale production systems.

The role of multinational corporations in driving international collaborations can't be overstated. Companies like ABB, Siemens, and Boston Dynamics are not only developing cutting-edge humanoid robots but are also actively participating in the creation of global standards. Through their involvement, they ensure that their innovations contribute to a cohesive, industry-wide framework that other manufacturers can adopt.

Another pivotal aspect of international collaboration is knowledge transfer through global educational programs and research initiatives. Universities and research institutes across the world collaborate on projects to push the boundaries of what AI-powered humanoid robots can achieve. These academic partnerships are crucial for training the next generation of engineers and researchers, equipping them with the skills necessary to advance the field further. For example, the EU's Horizon 2020 program has funded numerous collaborative projects that have led to significant advancements in robotic technology.

However, international collaboration isn't without its challenges. Differences in regulatory environments, business practices, and technological maturity levels may create roadblocks. Overcoming these challenges requires robust diplomatic engagement and a commitment to continuous dialogue and compromise.

Harmonizing regulatory frameworks across different countries is an ongoing challenge and an essential aspect of international collaboration. Countries must work together to align their regulations to avoid conflicts and redundancies that could stifle innovation. Collaboration among policymakers is crucial to formulating regulations that balance risk management with technological advancement.

The diversity of needs and priorities can make standardization efforts complex. Developing countries may prioritize cost-effective solutions, whereas developed nations might focus on cutting-edge technology. By addressing these varied requirements through collaborative research and discussions, a balanced approach can be achieved, ensuring that the benefits of robotic advancements are globally equitable.

Moreover, international collaboration brings ethical considerations to the forefront. Different countries may have varying perspectives on privacy, data security, and the role of robots in society. Tackling these ethical considerations through multinational forums ensures that the deployment of humanoid robots respects global norms and values.

International trade agreements also play a crucial role in the proliferation of robotics in a globalized economy. These agreements facilitate the free movement of robotic components and systems across borders, fostering a more integrated global market for advanced automation technologies.

Finally, examining successful case studies of international collaborations provides valuable insights. The partnership between South Korea and France in developing advanced robotic exoskeletons for industrial use cases stands out as a testament to what can be achieved through international cooperation. By leveraging South Korea's expertise in robotics and France's industrial know-how, they have managed to create solutions that are now employed globally, enhancing productivity and worker safety.

In summary, international collaborations and standardized frameworks are vital for the sustainable and ethical advancement of AI-powered humanoid robots in the globalized economy. These efforts not only ensure technological coherence and safety but also promote an environment of mutual growth and innovation. Through collective effort, the global community can harness the full potential of humanoid robots, revolutionizing industrial production and paving the way for a future where human and robotic workforces seamlessly coexist.

Case Studies from Different Regions

When exploring the transformative impact of AI-powered humanoid robots on the global manufacturing landscape, it's essential to look at specific implementations across different regions. Each region brings unique challenges and opportunities, highlighting the diverse ways these advanced machines are reshaping industrial settings. From Asia to Europe and the Americas, the adoption of humanoid robots varies significantly, driven by distinct economic, cultural, and regulatory factors. Let's delve into some compelling case studies that illustrate these regional differences.

In Japan, a country known for its technological prowess, the integration of humanoid robots into manufacturing has been a game changer. Japanese companies have been at the forefront of robotic innovation, with firms like FANUC and Kawasaki Heavy Industries pioneering the use of humanoid robots in assembly lines. These robots are not just increasing efficiency but are also addressing the issue of an aging workforce. In factories equipped with AI-powered robots, tasks that were once labor-intensive are now automated, allowing companies to maintain high production levels despite demographic challenges. This shift is particularly evident in the automotive sector, where humanoids are employed to perform precise and repetitive tasks, significantly reducing human error and enhancing product quality.

Moving to Europe, Germany offers another fascinating case study. Known for its robust manufacturing sector, particularly in engineering and automotive industries, Germany has leveraged AI and robotics to maintain its competitive edge. Companies such as BMW and Siemens have integrated humanoid robots into their production processes to work alongside human employees, fostering a collaborative environment. These robots are designed to assist with complex tasks that require a high degree of precision and adaptability. For instance, in BMW's production lines, humanoid robots are employed to assemble intricate parts, ensuring consistency and speed. This collaboration between humans and robots has not only improved productivity but also created a safer working environment by reducing the risk of workplace injuries.

In North America, the United States showcases a different approach to the adoption of humanoid robots. The focus here has been more on customization and versatility. American companies like Boston Dynamics and Rethink Robotics are developing humanoid robots tailored for specific industrial needs. One notable example is the use of humanoids in logistics and warehousing. Amazon, a leader in this space, utilizes a fleet of robots to streamline its fulfillment centers. These robots work alongside human workers, optimizing the picking and packing processes. What sets the US approach apart is the emphasis on data-driven decision-making. By leveraging big data analytics, companies can continually refine and improve the performance of their humanoid robots, ensuring they meet the dynamic requirements of various industries.

China, with its rapidly growing manufacturing sector, presents a unique case study in the deployment of humanoid robots. The Chinese government has been actively promoting automation to enhance productivity and reduce reliance on human labor. Companies like Foxconn, a major electronics manufacturer, have integrated tens of thousands of robots into their production lines. These robots are used for a wide range of tasks, from simple assembly to complex quality control processes. The result is a significant boost in production speed and a reduction in labor costs. Additionally, China's approach to humanoid robots includes substantial investment in AI research and development, aiming to produce more intelligent and autonomous machines. This strategy is part of a broader vision to position China as a global leader in AI and robotics.

India, another emerging player in the global manufacturing arena, is also turning to humanoid robots to overcome several challenges. The country faces issues related to labor costs and the need for increased efficiency in manufacturing processes. Indian companies are gradually adopting humanoid robots to perform monotonous and repetitive tasks that are prone to human error. An example can be seen in the textile industry, where robots are employed to handle fabric cutting and stitching, ensuring uniformity and reducing wastage. These efforts are supported by government initiatives aimed at promoting automation and innovation within the manufacturing sector.

Brazil, representing the Latin American region, offers insights into the challenges and opportunities of integrating humanoid robots in a developing market. The Brazilian manufacturing sector, particularly in the automotive and electronics industries, is slowly embracing automation to enhance productivity and compete globally. Companies like Embraer, a leading aerospace manufacturer, are exploring the use of humanoid robots in assembly lines to improve precision and reduce production times. However, the adoption of such advanced technologies faces hurdles, including high initial costs and the need for specialized skills. Addressing these challenges requires a concerted effort from both the private sector and the government, focusing on training and development programs to equip the workforce with the necessary skills.

Beyond these individual case studies, a common theme emerges: the global nature of humanoid robotics integration means that countries are learning from each other's successes and challenges. Collaborative international efforts, such as partnerships between academic institutions and industry leaders, are fostering innovation and accelerating the development and deployment of humanoid robots. For instance, European and Japanese firms have entered joint ventures to combine their expertise in robotics engineering and AI, leading to the creation of more advanced and capable machines. These cross-border collaborations ensure that the benefits of humanoid robots are realized on a global scale, driving productivity and economic growth.

Another noteworthy trend is the harmonization of regulatory standards to facilitate the smooth integration of humanoid robots across borders. Various international bodies and organizations are working towards establishing common safety and operational guidelines. This effort is crucial for ensuring that humanoid robots can operate seamlessly in different industrial environments without encountering regulatory hurdles. Countries participating in these initiatives gain a competitive edge by adopting best practices and ensuring that their industries remain at the cutting edge of technology.

Lastly, the societal impact of humanoid robots can't be overlooked. While regions like Japan and Germany have rapidly adapted to the presence of these robots in industrial settings, other regions are still grappling with public perception and acceptance. Cultural factors play a significant role in how humanoid robots are perceived and integrated. In regions where there is apprehension or resistance, educational campaigns and public engagement activities are essential to build trust and highlight the benefits of humanoid robotics. By addressing societal concerns and demonstrating the positive impacts on job quality and safety, regions can foster a more welcoming environment for these technological advancements.

In conclusion, the integration of AI-powered humanoid robots into manufacturing is a global phenomenon characterized by regional nuances. Whether it's Japan's response to an aging workforce, Germany's focus on human-robot collaboration, the United States' emphasis on customization and data-driven improvements, China's massive scale implementation, India's targeted adoption in specific industries, or Brazil's efforts to overcome adoption challenges, each case study offers valuable insights. These regional experiences not only highlight the transformative potential of humanoid robots but also underscore the importance of tailored approaches to ensure their successful integration. As the world continues to embrace these advanced technologies, the collective experiences from different regions will undoubtedly shape the future of industrial automation.

Chapter 18:
Humanizing Humanoids:
Social Robotics in Factories

As factories become increasingly automated, the role of humanoid robots in these environments is evolving from purely functional to more socially integrated. These advanced robots, equipped with emotional intelligence and sophisticated communication abilities, are transforming the industrial workspace by fostering more intuitive and effective human-robot interactions. This chapter delves into the development of robots that not only understand commands but can also interpret and respond to human emotions, creating a collaborative atmosphere where robots and humans work side by side seamlessly. By embedding social capabilities into robots, manufacturers aim to enhance teamwork, reduce friction, and ultimately boost productivity. The intersection of AI with social robotics is not just a technological milestone but a revolutionary shift in how we perceive co-working environments, making the factories of the future more human-friendly and efficient.

Building Emotional Intelligence in Robots

In the rapidly evolving landscape of social robotics within factories, instilling emotional intelligence in humanoid robots marks a significant milestone. Emotional intelligence, once considered an exclusively human trait, is now becoming an integral aspect of robot development. This transformation is driven by the need for seamless human-robot interaction, fostering cooperation, and enhancing productivity. Emotional intelligence in robots involves the nuanced recognition, understanding, and appropriate response to human emotions, promoting a more collaborative and efficient working environment.

One of the key components of building emotional intelligence in robots is the development of advanced algorithms that can process and interpret human emotions. These algorithms leverage machine learning and artificial intelligence to analyze facial expressions, voice tones, and body language. By integrating these capabilities, robots can respond to the emotional states of their human counterparts in real time, fostering smoother interactions. For instance, a robot detecting stress in an assembly line worker might adjust its approach, offering assistance or reducing its pace to match the worker's comfort level.

Emotional intelligence in robots isn't merely about identifying and reacting to emotions. It's about creating a feedback loop where robots learn from each interaction. Over time, they develop a deeper understanding of human behaviors and preferences. This iterative learning process is critical in refining the emotional acuity of robots. Through continuous exposure and adaptation, robots can better predict human needs and respond more naturally, ultimately creating a more harmonious working environment.

However, the journey to developing emotionally intelligent robots isn't without its challenges. One significant hurdle lies in the diversity of human emotions and expressions. Unlike predefined tasks or routines that robots execute with high precision, human emotions are complex, varied, and context-dependent. To address this, researchers are employing vast datasets of emotional cues and contextual information to train robots. These datasets encompass a wide range of cultures, age groups, and professional environments to ensure that the robots' emotional intelligence is comprehensive and adaptable.

Another important factor is the integration of multimodal sensory inputs. Humanoid robots are equipped with a combination of visual, auditory, and tactile sensors, each contributing to the overall emotional understanding. For example, visual sensors capture facial expressions, auditory sensors analyze tone and pitch of voice, and tactile sensors gauge physical interactions. Merging these sensory inputs allows for a more holistic perception of human emotions, leading to more accurate and empathetic responses from the robots.

Moreover, it is essential to consider the ethical implications of emotionally intelligent robots. These robots will constantly collect and analyze personal data related to employees' emotional states. Ensuring the privacy and security of this data is paramount. Industries deploying such robots must adhere to strict data protection regulations and implement robust security measures to prevent misuse of sensitive information. Transparency in how emotional data is collected and used will also play a crucial role in gaining employee trust and acceptance.

The benefits of building emotional intelligence in robots extend beyond improving human-robot interactions. Emotionally intelligent robots can contribute to a positive workplace culture by identifying and mitigating stress or burnout among employees. For instance, they can alert supervisors when an employee exhibits signs of prolonged distress, prompting timely interventions. This proactive approach can enhance overall employee well-being and productivity, creating a more supportive and collaborative factory environment.

On the technological front, integrating cloud-based solutions and edge AI further augment the emotional capabilities of robots. Cloud computing facilitates the processing and analysis of extensive emotional data, while edge AI enables real-time responses. This combination ensures that robots remain responsive and adaptable, continuously improving their emotional intelligence based on real-time feedback and historical data.

Furthermore, the role of natural language processing (NLP) cannot be overstated in this context. NLP enables robots to interpret and generate human language with emotional nuance, enhancing verbal communication. Robots equipped with advanced NLP can engage in more meaningful conversations, offering comfort, encouragement, or pertinent information as needed. This not only makes robots more relatable but also more effective in their roles within the factory.

Collaborative initiatives between technology developers, psychologists, and industrial engineers are driving the evolution of emotional intelligence in robots. Psychologists provide insights into human emotional responses, which are vital for developing accurate and empathetic AI models. Industrial engineers ensure that these models are practically applicable within the factory setting, balancing emotional intelligence with operational efficiency. This interdisciplinary approach is crucial for creating emotionally intelligent robots that are both technically sophisticated and attuned to the human experience.

In conclusion, building emotional intelligence in robots represents a groundbreaking step in the field of social robotics within industrial settings. As robots become more adept at recognizing and responding to human emotions, they will foster a more collaborative, efficient, and emotionally supportive work environment. The intersection of advanced AI, comprehensive emotional datasets, and multimodal sensory inputs holds the key to this transformation. While challenges remain, the ongoing collaboration between various disciplines promises a future where emotionally intelligent robots are an integral part of the industrial workforce, enhancing both productivity and employee well-being.

Enhancing Communication and Interaction

Communication and interaction sit at the heart of effective collaboration, and it's no different when discussing the integration of humanoid robots into factory settings. By enhancing these elements, not only do the robots become more efficient in their roles, but they also facilitate smoother workflows and contribute to a more harmonious working environment. For AI-powered humanoid robots to truly achieve their potential, it's imperative that they engage with human workers effectively and intuitively.

To reach that level of interaction, one foundational aspect is natural language processing (NLP). NLP allows humanoid robots to understand and respond to human speech in ways that are contextually appropriate and constructive. Advanced NLP models enable robots to grasp nuanced language cues, such as tone and sarcasm, which can significantly impact workplace communication. An employee needing assistance shouldn't have to code their query into robot-speak; instead, the robot should understand natural human language.

Consider a factory setting where a line worker might ask a robot, "Can you give me the torque wrench from the toolbox?" An AI-powered humanoid, equipped with advanced NLP, can identify the specific tool from a list of potential items and understand the urgency and context of the request, thereby enhancing operational efficiency.

While voice commands are a crucial component, visual cues also play an essential role in communication. Humanoid robots equipped with state-of-the-art visual recognition systems can interpret gestures, facial expressions, and body language. This capability makes interactions more fluid and lessens the need for repetitive verbal commands. In a bustling factory floor, being able to understand a nod or a gesture can shave precious seconds off interactions and make the entire production line more responsive.

Augmented reality (AR) is another exciting frontier in enhancing robot-human communication. Imagine a factory setting where both humans and robots can see shared information overlays through AR displays. This synchronous data interaction can guide both parties through complex tasks, instantly pointing out potential errors or optimizations. By visualizing the same data, humans and robots can achieve a higher level of collaboration.

Another critical component is emotional intelligence. As futuristic as it may sound, integrating emotional intelligence into humanoid robots can provide them with the empathy and adaptability required for more personalized interactions. Emotionally intelligent robots can adjust their behavior based on the emotional states of the human workers around them. If a robot detects that a worker is stressed or frustrated, it can modify its interactions accordingly, perhaps offering a calmer voice tone or providing additional assistance. Achieving this level of interaction deepens the trust between human and robotic colleagues.

For example, a robot that notices a worker's frustration might say, "It looks like you're having a tough time. Would you like some help?" Imagine the boost in productivity and morale if workers felt their automated counterparts were genuinely supportive. This emotional attunement goes beyond mere task execution; it humanizes the workplace, making robots seem like empathetic partners rather than cold, impersonal machines.

Integrating these communicative capabilities isn't without its challenges. Data privacy and security become paramount when robots process personal and potentially sensitive information. Ensuring that all interactions are encrypted and that personal data is never misused is a critical responsibility for designers and operators. The balance between functional benefit and ethical practice must always be maintained.

Beyond ethical considerations, training for both robots and humans is crucial. Robots must be continually updated with new language models and visual recognition databases, while human workers need training sessions to feel comfortable and proficient in interacting with their robotic colleagues. This dual approach ensures both sides of the equation are optimized for effective collaboration.

To further interaction enhancement, feedback loops are essential. Humanoid robots should constantly receive and act upon feedback from their human coworkers. Did a worker find a command ambiguous or unhelpful? The robot can adapt future actions based on this input. Over time, this iterative process leads to highly personalized and efficient cooperation.

Improved interaction also opens the door for robots to take on more diverse roles within factories. From customer interaction roles to maintenance and beyond, a robot that can communicate clearly and effectively is immensely versatile. This versatility allows companies to deploy these robots in multiple areas, maximizing their return on investment.

Effective communication isn't just about robots understanding humans; it's also about humans understanding robots. User-friendly interfaces and prompt systems should offer clear instructions and feedback to workers, demystifying robotic operations and reducing apprehensiveness. Robots equipped with visual screens can provide real-time data and analytics, helping workers make informed decisions quickly.

Collaborative robots that can exchange ideas with humans—and even other robots—can create a dynamic and engaging work environment. For instance, a robot might flag inefficiencies in real-time and suggest immediate improvements, rather than simply following a pre-programmed task sequence. Workers, in turn, could offer suggestions or report issues, leading to a productive ebb and flow of communication.

Think of a scenario where a humanoid robot in a factory suggests an optimized route for product assembly based on real-time data analysis, while a human worker notices a potential troubleshooting area that the robot didn't flag. This teamwork not only raises efficiency but fosters a culture of continuous improvement and shared responsibility.

Additionally, multi-lingual capabilities can break down language barriers in diverse workplaces. A humanoid robot that understands and communicates in several languages facilitates inclusion and efficiency among multicultural workgroups. This linguistic flexibility can be particularly impactful in global manufacturing sites.

In conclusion, enhancing communication and interaction between humanoid robots and human workers isn't just a matter of technology; it's about creating an environment where both can thrive harmoniously. By leveraging advanced NLP, visual recognition, AR, emotional intelligence, and feedback systems, robots in factories can become not just tools but collaborative partners. This relationship can significantly increase efficiency, improve morale, and lay the groundwork for the future of industrial automation.

Chapter 19:
The Supply Chain Transformation

The supply chain transformation, driven by AI-powered humanoid robots, is revolutionizing how industries approach the journey from raw materials to product delivery. With intelligent robotics seamlessly integrated into every stage, from procurement and logistics to manufacturing and distribution, companies are experiencing unprecedented efficiency. These robots, armed with advanced sensors and real-time data processing capabilities, significantly reduce downtime by predicting maintenance needs and autonomously adjusting workflows. The result is a more resilient, adaptive supply chain that not only meets today's demands but is also flexible enough to adapt to tomorrow's challenges. This sophisticated synergy of AI and robotics is setting new standards for productivity, sustainability, and innovation in modern manufacturing landscapes.

From Raw Materials to Product Delivery

In the world of traditional manufacturing, the journey from raw materials to the final product involves numerous steps, each of which is critical for ensuring efficiency and quality. With the advent of AI-powered humanoid robots, this journey is evolving into a streamlined, more intelligent process. The integration of these robots into various stages of the supply chain is revolutionizing how products are conceived, designed, manufactured, and delivered.

Humanoid robots equipped with artificial intelligence are capable of performing complex tasks that were once the domain of human workers. From the assembly line to quality control, these robots are not only replicating but also enhancing human capabilities. Their ability to work with precision, speed, and consistency has significant implications for the entire supply chain.

Consider the initial stages of production, where raw materials are procured and prepared for manufacturing. In a traditional setting, this requires human oversight to ensure that the materials meet stringent quality standards. AI-powered humanoid robots step in to automate this process, using advanced sensor technology to assess and categorize materials with unprecedented accuracy. This minimizes human error, reduces waste, and ensures a higher standard of raw material quality.

Once the raw materials are ready, the focus shifts to the assembly line. Here, humanoid robots bring in their strength: the ability to perform repetitive tasks with precision and without fatigue. Unlike their human counterparts, these robots can work around the clock, increasing productivity and reducing the time it takes to move from raw material to finished product. They can also adapt to different product lines with minimal reprogramming, making them invaluable for manufacturers who need to switch between various products rapidly.

One of the most transformative aspects of AI-powered humanoid robots is their contribution to quality control. Traditional quality control methods require extensive human resource involvement, which can be both time-consuming and prone to error. However, humanoid robots equipped with machine learning algorithms can inspect products in real-time, identifying defects at a scale and speed beyond human capacity. This ensures that only the highest quality products make it through the final stages of the supply chain.

After the products pass quality control, the journey continues to packaging and logistics. In the packaging stage, humanoid robots introduce a level of efficiency that is hard to match. They can handle delicate items with care and package them in a way that maximizes space and minimizes shipping costs. Furthermore, these robots can integrate seamlessly with warehouse management systems, optimizing inventory management and ensuring that products are stored and tracked efficiently.

When it comes to logistics, humanoid robots are making strides in addressing some of the most significant challenges faced by the industry. AI algorithms enable these robots to plan the most efficient delivery routes, taking into account variables such as traffic, weather conditions, and delivery priorities. This level of logistical intelligence not only speeds up delivery times but also reduces fuel consumption and environmental impact, aligning with broader sustainability goals.

Imagine a scenario where a factory is producing high-demand consumer electronics. The supply chain begins with suppliers sending raw materials, which humanoid robots inspect and process at the factory's docks. These robots are equipped with sensors capable of detecting impurities or defects in materials, ensuring that only premium components enter the manufacturing cycle. This level of vigilance in the initial stages sets the foundation for a seamless production flow.

Next, the materials move to the assembly line, where different types of humanoid robots specialize in tasks ranging from intricate circuit board assembly to the more physical labor of chassis construction. Each robot is programmed to execute its role with precision, ensuring congruity across the board. The assembly process becomes a dance of synchronized robotic movements, each contributing to the creation of the final product.

With the assembly complete, another set of humanoid robots takes over for quality assurance. These robots, equipped with advanced vision systems, inspect each unit in detail. They detect flaws that might be missed by the human eye, such as micro-cracks or misaligned components. Any unit that doesn't meet the stringent quality standards is flagged for rework or recycling, ensuring customers receive nothing but the best.

The final step in the factory phase involves packaging. Humanoid robots equipped with AI capabilities determine the optimal way to pack products, considering factors like size, shape, and fragility. These robots can adapt on the fly, changing packing strategies based on real-time data. This adaptability ensures that products are packed securely and efficiently, reducing the risk of damage during transit.

After packaging, the logistical journey begins. Humanoid robots, aided by predictive analytics, optimize the loading of goods onto delivery vehicles, making sure every inch of space is utilized effectively. They generate the most efficient delivery routes, minimizing delays and expediting the entire logistics process. These robots also use real-time data to navigate obstacles, ensuring that products reach their destination on time.

One noteworthy example of how AI-powered humanoid robots have revolutionized the supply chain is in the automotive industry. Here, the implementation of these robots has drastically condensed the time from raw materials to product delivery. In the past, manufacturing a single car required a multitude of human tasks, each susceptible to delays and errors. Today, humanoid robots handle everything from welding and painting to final assembly and inspection, dramatically accelerating production timelines and boosting overall efficiency.

This enhanced efficiency translates not only to faster production times but also to a reduction in operational costs. Factories can now produce more goods in less time and with fewer resources. By reducing the margin of error and the need for rework, companies are saving significant amounts of money. These cost savings can be passed down to consumers, making high-quality products more affordable and accessible.

In addition to operational improvements, humanoid robots contribute to a safer working environment. By taking over dangerous and repetitive tasks, they reduce the risk of injury to human workers. This shift not only improves workforce morale but also lowers the costs associated with workplace injuries and insurance premiums.

Looking ahead, the role of humanoid robots in the supply chain will only expand. As AI and machine learning algorithms continue to advance, these robots will become even more adept at handling complex tasks and making autonomous decisions. This anticipated growth will further refine the efficiency of the supply chain, from raw materials to product delivery, paving the way for innovations we can only begin to imagine.

Reducing Downtime with Intelligent Robotics

In the age of high-speed manufacturing, minimizing downtime is crucial for maintaining operational efficiency. Unplanned downtime can cause substantial financial losses and disrupt supply chain continuity. As we venture deeper into "The Supply Chain Transformation," one of the most groundbreaking developments is the integration of AI-powered humanoid robots. These intelligent systems offer innovative solutions to reduce downtime, thereby considerably enhancing productivity and reliability.

AI-powered humanoid robots stand out for their advanced diagnostic and predictive maintenance capabilities. Traditional industrial robots have long served well for repetitive tasks, but they lack the capability to anticipate and resolve issues autonomously. In contrast, AI-enabled robots can monitor their own performance and alert human supervisors to potential issues before they cause downtime. Imagine a robot in a production line detecting a subtle decrease in motor efficiency. It can alert the maintenance team, who can address the issue immediately, preventing costly interruptions.

The significance of predictive maintenance cannot be overstated. AI-powered robots use a vast array of sensors to gather data and machine learning algorithms to analyze patterns. These patterns help predict when a component is likely to fail, allowing for timely maintenance or replacement. For example, sensors can monitor temperature, vibration, and other critical parameters in real-time. When deviations from the norm are detected, the system flags the issue for further inspection. This approach is far more efficient than scheduled maintenance, which can result in unnecessary downtime and costs.

Additionally, humanoid robots equipped with advanced AI can conduct self-repair to an extent. This doesn't mean complete autonomy, but the robot can handle simple fixes, such as recalibrating a sensor or updating its own software to resolve minor glitches. These self-repair capabilities substantially reduce the downtime that typically accompanies waiting for human intervention, allowing factories to maintain near-constant operation.

Another important aspect of reducing downtime is the ability of intelligent robots to collaborate with humans effectively. These robots are designed to work alongside their human counterparts, filling in gaps when humans are unavailable or augmenting human skills with their precision and strength. For example, if a machine requires a part that a worker needs to fetch from storage, an intelligent robot might quickly accomplish this task, ensuring the production line keeps moving.

Today's factory floors are highly intricate environments, demanding constant monitoring and quick decision-making. The deployment of intelligent robotics ensures that these environments maintain their operational rhythm. Robots can take on multiple roles within a plant, from performing quality checks to facilitating complex assembly tasks. Imagine a scenario where part of the line faces a temporary glitch; a humanoid robot can be rerouted to another part of the operation where its capabilities are still needed, thus minimizing impact across the board.

Downtime reduction also comes from more strategic planning and logistics facilitated by AI. Intelligent robots don't just perform tasks; they also process vast amounts of data to help optimize supply chain operations. AI can analyze workflows, identify bottlenecks, and recommend adjustments. Those insights facilitate more fluid operations and help in preemptively solving issues that could lead to downtime.

Moreover, the integration of cloud computing with AI-powered robotics has broadened the horizon for information sharing and real-time updates. Humanoid robots can access up-to-date information on parts, machine conditions, and operational status. This connectivity ensures that a factory's entire network of robots and machines operates cohesively, and any potential problem is swiftly communicated and rectified, reducing the risk of extended downtime.

Real-life case studies have shown the incredible potential of intelligent robotics in reducing downtime. Early adopters report not only decreased unplanned downtime but also improved overall efficiency and output quality. For example, major automotive manufacturers using AI-powered robots have seen production lines run longer with fewer interruptions, resulting in considerable cost savings and higher product throughput.

Integrating intelligent robotics can also lead to the optimization of human labor. By reallocating mundane, repetitive tasks to robots, human workers can focus on more critical, cognitively demanding tasks. This maximizes human skill sets while ensuring that the production line remains efficient and less prone to stoppages. The human workforce becomes more strategic, leading problem-solving efforts while robots handle predictable operations.

Another key benefit against downtime comes from the learning capabilities of AI-powered robots. These systems don't just operate based on pre-loaded algorithms; they continuously learn from their environment and adapt to changing conditions. Machine learning enables robots to become more efficient over time, reducing the frequency and length of downtime. Every operational hour becomes a lesson that the robot learns, making it more adept at foreseeing and mitigating issues.

Overall, integrating intelligent robotics within the supply chain is a game-changer. Reducing downtime through AI-powered humanoid robots is a multifaceted process involving advanced diagnostics, predictive maintenance, self-repair capabilities, enhanced human-robot collaboration, data analytics, and machine learning. The result is not just a more resilient and efficient supply chain but also a paradigm shift in how we perceive modern manufacturing.

As we look ahead, the ongoing advancements in AI and robotics will only amplify these benefits. Future robots will be even more autonomous, reliable, and intelligent, pushing the limits of what's possible in downtime reduction. This progression underscores the importance of AI in the ongoing transformation of the supply chain, heralding a new era of efficiency and reliability in industrial production.

Chapter 20:
The Impact of AI on Innovation

A I is not just propelling technological advancement; it's redefining the entire landscape of innovation across industries. The introduction of AI-powered humanoid robots into factory settings is accelerating research and development processes by enabling rapid prototyping and intelligent design iteration. These robots are not just tools; they're collaborative partners that foster a culture of continuous improvement, driving efficiency and creativity. By leveraging machine learning algorithms and vast datasets, manufacturers can now anticipate trends, optimize production workflows, and make data-driven decisions with unprecedented precision. This transformative impact of AI is breaking down traditional barriers, making innovation more accessible and sustainable, and setting new benchmarks for productivity and quality, thus reshaping the future of industrial automation in profound ways.

Accelerating Research and Development

Artificial intelligence (AI) has long been heralded as the next big thing in technology, with its potential to revolutionize industries and create new paradigms. Nowhere is this more evident than in the field of research and development (R&D). In the pursuit of innovation, AI-powered humanoid robots have become indispensable partners. By bolstering the capabilities of R&D teams, these robots have accelerated the pace of discovery and optimization, leading to breakthroughs that were once considered science fiction.

One of the most profound impacts of AI on R&D is the sheer speed at which it can process and analyze data. Traditional methods, hampered by limited computational power and manual labor, often resulted in lengthy timelines for concept validation and product development. In contrast, AI algorithms can sift through vast datasets in a fraction of the time, identifying patterns and insights that might take human researchers years to uncover. This accelerates the initial stages of R&D, enabling quicker iterations and refinements.

Humanoid robots equipped with advanced machine learning capabilities are particularly transformative. These robots can perform repetitive tasks with precision, freeing up human researchers for more complex and creative problem-solving. Additionally, their ability to learn from each iteration ensures continuous improvement in processes. For instance, in the pharmaceutical industry, AI-powered humanoid robots play pivotal roles in drug discovery. By simulating molecular interactions, these robots can predict the efficacy and safety of new compounds, drastically reducing the time it takes to bring new drugs to market.

Moreover, the integration of AI in R&D fosters a multidisciplinary approach. Traditionally, engineering, design, and testing stages were siloed, with each team working in isolation. AI breaks down these barriers by facilitating seamless communication and collaboration among different teams. Through shared platforms and real-time data analytics, humanoid robots help create harmonious workflows where each stage of R&D informs the next, streamlining the entire process.

The impact isn't limited to just speeding up processes and enhancing collaboration. AI-powered humanoid robots also bring a level of precision and accuracy that is difficult to achieve through human effort alone. This is especially critical in fields that require meticulous attention to detail, such as aerospace engineering and biotechnology. With their advanced perception and manipulation skills, humanoid robots can perform delicate operations, measure variables with exacting precision, and ensure that prototypes adhere to the strictest standards.

AI's role in accelerating R&D goes beyond operational efficiency; it also fosters a culture of continuous improvement. Humanoid robots equipped with AI are not just static tools; they are dynamic, learning entities that evolve with each task they perform. This continuous learning loop means that each R&D cycle benefits from the lessons of the previous ones, creating an environment where incremental improvements lead to substantial progress over time. The iterative learning process cultivates an ethos of perpetual innovation, encouraging human researchers to push the boundaries of what's possible.

Additionally, AI-powered humanoid robots enhance predictive capabilities within R&D teams. By leveraging machine learning models, these robots can forecast potential challenges and outcomes, enabling researchers to proactively address issues before they become critical. Predictive analytics can inform decision-making processes, optimizing resource allocation and minimizing wasted effort. This foresight is invaluable in industries where time and resources are inextricably linked to competitive advantage.

Another significant benefit is the democratization of R&D capabilities. Smaller companies and startups, which traditionally lacked the resources to compete with larger organizations, can now leverage AI-powered humanoid robots to level the playing field. Advanced R&D tools and methodologies that were once the exclusive domain of well-funded enterprises are becoming accessible to a broader spectrum of innovators. This democratization spurs a new wave of creativity and ingenuity, as diverse voices contribute to the collective advancement of technology.

Furthermore, the ethical considerations of deploying AI in R&D cannot be overlooked. While the promise of accelerated innovation is enticing, it is crucial to ensure that the development and deployment of AI technologies are guided by ethical principles. Humanoid robots in R&D must operate within frameworks that prioritize fairness, transparency, and accountability. This ensures that the benefits of accelerated research are equitably distributed and that potential risks are managed responsibly.

The convergence of AI and humanoid robotics in R&D also has far-reaching implications for education and workforce development. As AI-powered tools become integral to R&D processes, there is a growing need for skilled professionals who can navigate this new landscape. Educational institutions and industry leaders must collaborate to develop curricula and training programs that equip the next generation of researchers with the skills needed to harness the full potential of AI. This includes not only technical proficiency but also an understanding of the ethical and societal implications of AI in R&D.

The integration of AI in R&D is reshaping the global competitive landscape. Countries and companies that invest in AI-powered R&D infrastructure are positioned to lead in innovation, gaining a strategic advantage in the global market. This shift underscores the importance of comprehensive AI strategies at both the organizational and national levels. Policymakers and industry leaders must work together to create ecosystems that support sustained investment in AI and R&D, driving long-term economic growth and technological leadership.

In conclusion, the impact of AI on research and development is transformative, ushering in a new era of accelerated innovation. AI-powered humanoid robots enhance speed, precision, collaboration, and predictability in R&D processes, driving breakthroughs across various industries. As we embrace these advancements, it is essential to remain mindful of ethical considerations and to invest in education and infrastructure that support the responsible and equitable deployment of AI. With the right strategies in place, the fusion of AI and R&D holds the promise of a future where technological progress is both rapid and inclusive.

Fostering a Culture of Continuous Improvement

Industries that have integrated AI-powered humanoid robots are paradigms of innovation, and one of the cornerstones of their success is a culture of continuous improvement. This culture isn't just a set of practices; it's a mindset that permeates every level of operation, encouraging constant growth, learning, and adaptation. The relentless pursuit of better performance and efficiency keeps industries ahead of the curve, driving the revolutionary changes AI promises.

For starters, fostering a culture of continuous improvement involves a systematic approach to identifying areas for enhancement. In an environment where AI-driven humanoid robots work alongside humans, data plays a pivotal role. Sensor technologies and AI algorithms provide a steady stream of insights into performance metrics, operational bottlenecks, and potential points of failure. By leveraging these insights, organizations can continuously refine their processes, ensuring that both human and robotic workers are operating at peak efficiency.

To embed this culture deeply, it's essential to promote an environment that encourages feedback and innovation. Employees on the factory floor should feel empowered to suggest improvements and experiment with new approaches. Similarly, routine performance evaluations of humanoid robots should be conducted to identify potential upgrades in software and hardware, driving the evolution of these machines. When both human and robot teams are invested in the process of continuous refinement, the results can be astounding, leading to faster production cycles, fewer errors, and more innovative solutions to complex problems.

Training and ongoing education are also crucial elements of a culture of continuous improvement. As humanoid robots become more sophisticated, the need for employees to possess up-to-date knowledge and skills becomes paramount. Companies must invest in robust training programs that keep their workforce proficient in operating and interacting with these advanced systems. Additionally, cross-disciplinary teams that include AI specialists, engineers, and human operators can foster a collaborative environment where knowledge sharing is the norm.

Another aspect of fostering this culture is the adoption of iterative development cycles. Traditional approaches to manufacturing often relied on linear processes, but in the age of AI, iteration is key. Continuous improvement thrives in a setting where changes can be rapidly tested, evaluated, and iterated upon. Implementing agile methodologies, commonly used in software development, can yield significant benefits in industrial settings. This approach ensures that enhancements are made based on real-time data and immediate feedback rather than prolonged, less frequent evaluations.

Moreover, leadership plays a significant role in cultivating this mindset. Leaders need to set an example by embracing change and encouraging a forward-thinking approach. When management demonstrates a commitment to continuous improvement, it sends a powerful message throughout the organization. Regularly scheduled meetings that focus solely on innovation and improvement help keep these goals at the forefront of industrial operations. It's in these settings that the next big idea might emerge—a new algorithm to enhance robotic vision, a tweak in the production line that saves hours, or a novel approach to human-robot collaboration.

Despite the technological advances, human ingenuity remains at the heart of continuous improvement. AI-powered humanoid robots, while powerful, are tools that augment human potential. They provide the data and capabilities, but it's the human teams that interpret this information, make strategic decisions, and drive progress. Companies that recognize this symbiosis and invest in both their workforce and their robotic assets stand to gain the most.

In fostering a culture of continuous improvement, companies must also remain vigilant about complacency. Even when systems are performing well, there's always room for growth. Disruption comes swiftly in today's technology-driven world, and what works today might be obsolete tomorrow. Thus, maintaining a relentless focus on improvement—not just in manufacturing processes but also in the capabilities of humanoid robots themselves—is essential.

Additionally, fostering continuous improvement means embracing failure as a learning opportunity. In a culture that celebrates innovation, mistakes are not seen as setbacks but as valuable data points. This perspective is vital in an ecosystem where both AI and human actors must navigate a complex and ever-changing landscape. Encouraging calculated risks can lead to breakthroughs that redefine what's possible in industrial automation.

Lastly, partnerships and collaborations with academia, technology developers, and other industries can significantly enhance a company's ability to foster continuous improvement. These partnerships can introduce new technologies, methodologies, and perspectives that might not be apparent within the confines of a single organization. Collaborative ventures offer a fertile ground for sharing best practices, co-developing cutting-edge solutions, and staying at the forefront of technological advancements.

Charlie Addison

In conclusion, the impact of AI on innovation is profound, but maximizing this impact requires a deliberate and sustained effort to foster a culture of continuous improvement. This involves leveraging data, empowering employees, promoting iterative development, committing to ongoing training, displaying strong leadership, valuing human ingenuity, avoiding complacency, embracing failure, and fostering partnerships. When these elements are in place, industries can unlock the true potential of AI-powered humanoid robots, driving unprecedented levels of efficiency, innovation, and success.

Chapter 21:
The Role of Cloud Computing and Edge AI

Cloud computing and Edge AI are pivotal in pushing the boundaries of what's possible with AI-powered humanoid robots in industrial settings. By leveraging the immense processing power and scalability of the cloud, factories can manage and analyze vast amounts of data from numerous robots in real time. This seamless connectivity ensures continuous learning and optimization, enhancing operational efficiency and reducing downtime. On the other hand, Edge AI enables robots to process data locally, making immediate decisions without latency. This is crucial for tasks requiring rapid response and high precision, such as quality control or real-time predictive maintenance. Together, these technologies create a robust infrastructural backbone that allows humanoid robots to perform complex functions with astonishing speed and accuracy, revolutionizing the landscape of industrial automation.

Enhancing Capabilities with Cloud Solutions

Cloud computing, a cornerstone of modern technology, plays an indispensable role in the evolution of AI-powered humanoid robots in industrial environments. By leveraging cloud solutions, factories can harness unprecedented processing power, storage, and scalability that were previously unattainable. This, in turn, enhances the operational capabilities of humanoid robots, paving the way for smarter and more efficient production processes. Imagine a scenario where robots no longer need to rely solely on their onboard processors but can tap into vast computational resources spread across the globe. This is the reality that cloud computing enables, dramatically transforming the efficiency and efficacy of industrial robotics.

One of the most substantial benefits of integrating cloud computing with humanoid robots is the ability to process and analyze massive datasets quickly and efficiently. In a factory setting, data is continually generated from various sources—sensors, machinery, and even the robots themselves. Cloud platforms can rapidly process this influx of data, providing real-time analytics and actionable insights. Consequently, the robots can make more informed decisions, optimizing production lines and reducing downtime. This combination of cloud computing and robotics creates a symbiotic relationship where machines learn and adapt, enhancing overall operational efficiency.

Cloud solutions also offer unparalleled flexibility in terms of software and updates. In traditional setups, updating a robot's software could be a cumbersome and time-consuming process. With cloud-based architectures, updates can be deployed remotely and instantaneously across an entire fleet of robots. This ensures that all units are running the latest algorithms and security patches, minimizing vulnerabilities and maximizing performance. Moreover, it allows for the continuous improvement of AI models, as developers can push upgrades and refinements at any time, keeping the robots at the cutting edge of technology.

Scalability is another crucial advantage that cloud computing brings to the table. Factories operating on a larger scale require significant computational resources, especially when dealing with multiple robots performing complex tasks. Cloud platforms provide on-demand scalability, allowing factories to easily scale their computing resources up or down based on their needs. This ensures that the infrastructure can handle peak loads without compromising performance. As a result, factories can maintain high productivity levels without the need for substantial upfront investments in hardware.

Furthermore, cloud computing facilitates global collaboration and data sharing. In the context of humanoid robots, this means that researchers, engineers, and developers from different parts of the world can collaborate seamlessly, sharing data, models, and insights. Such a collaborative environment accelerates innovation and fosters the development of more sophisticated and capable robotic systems. This also opens the door for more standardized solutions across different industries, making it easier for companies to adopt and integrate humanoid robots into their operations.

Security is a paramount concern in any industrial setting, and cloud solutions offer robust measures to protect sensitive data. Advanced encryption techniques, secure access protocols, and regular security audits are just a few examples of how cloud providers ensure data integrity and privacy. While concerns about data breaches exist, cloud providers invest heavily in securing their infrastructure, often surpassing the security capabilities of in-house systems. This reassures organizations that their operational data and intellectual property are safeguarded against unauthorized access and cyber threats.

An often overlooked, but equally important benefit of cloud computing is cost efficiency. By moving computational tasks to the cloud, factories can reduce their reliance on expensive on-premises hardware, thereby lowering capital expenditures. Operational costs are also minimized as maintenance, and upgrades are managed by the cloud service providers. This cost-effectiveness allows even small and medium-sized enterprises to leverage advanced AI technologies, democratizing access to powerful tools that were once the preserve of large corporations.

Moreover, cloud solutions enable extensive data storage capabilities. Factories generate vast amounts of data daily, ranging from production metrics to maintenance logs. Storing this data effectively is crucial for ongoing analysis and future reference. Cloud platforms provide scalable storage solutions that can handle this data deluge, ensuring that valuable information is not lost. This historical data can be analyzed to identify trends, forecast future demands, and inform strategic decision-making, further enhancing the operational intelligence of humanoid robots.

In addition to enhancing operational efficiency, cloud computing also supports the development and deployment of advanced simulation and training environments. Factories can create virtual replicas of their physical setups, commonly known as digital twins, hosted on cloud platforms. These digital twins allow for extensive testing and optimization of robotic processes without disrupting actual production lines. Robots can be trained in these simulated environments, learning to navigate and perform tasks before being deployed in the real world. This not only reduces the risk of errors but also accelerates the onboarding process, allowing new robots to become productive members of the workforce more quickly.

Interoperability is another significant advantage offered by cloud computing. In a dynamic industrial environment, multiple systems, and devices need to communicate seamlessly. Cloud platforms act as a centralized hub, facilitating smooth data exchange and integration across various systems. This interoperability ensures that humanoid robots can work in harmony with existing machinery, sensors, and enterprise systems, streamlining workflows and enhancing overall productivity. By breaking down silos and fostering connectivity, cloud computing creates a more cohesive and efficient industrial ecosystem.

The ability to perform predictive maintenance is one of the standout features enabled by cloud solutions. Through continuous monitoring and analysis of data collected from robotic systems, cloud platforms can predict potential failures and maintenance needs before they occur. This proactive approach minimizes downtime and extends the lifespan of robotic systems, ensuring that they operate at peak efficiency. By leveraging predictive analytics, factories can schedule maintenance activities more effectively, reducing the risk of unexpected breakdowns and costly interruptions.

Another exciting prospect of cloud computing in the realm of humanoid robotics is the potential for real-time collaboration and task sharing. Robots across different locations can coordinate their activities through cloud-based platforms, allowing for synchronized operations and efficient task distribution. For instance, if a particular robot encounters a complex task it can't handle, it can seamlessly delegate the task to a more capable unit, irrespective of its location. This level of coordination and flexibility enhances the overall productivity of the workforce, paving the way for more intelligent and responsive production systems.

Lastly, cloud computing fosters a culture of continuous improvement and innovation. The ability to collect, store, and analyze vast amounts of data empowers organizations to refine their processes continually. AI models can be updated and improved based on real-world performance data, ensuring that the robots remain adaptive and responsive to evolving challenges. This iterative process of refinement and enhancement drives technological advancement, making humanoid robots smarter, more efficient, and more capable over time.

In conclusion, the integration of cloud computing solutions significantly enhances the capabilities of AI-powered humanoid robots in industrial settings. From real-time data processing and scalability to cost efficiency and predictive maintenance, the benefits are manifold. By leveraging the power of the cloud, factories can optimize their operations, drive innovation, and maintain a competitive edge in an increasingly automated and data-driven world. The future of industrial automation is undoubtedly intertwined with the advancements in cloud technology, propelling humanoid robots to new heights of efficiency, intelligence, and collaboration.

Edge Analytics for Real-Time Decision Making

In the landscape of modern manufacturing, the term "edge analytics" has become increasingly significant, particularly with the rise of AI-powered humanoid robots. Edge analytics refers to the processing of data at the source of data generation, minimizing latency and providing immediate insights. In contrast to traditional data analytics, which processes data in centralized cloud servers, edge analytics operates on local devices. This shift is crucial for real-time decision making, a necessity in the fast-paced, precision-driven environments of factories.

The integration of edge analytics allows humanoid robots to make instantaneous decisions, vital for maintaining smooth operations on the factory floor. These robots process data from an array of sensors—detecting everything from temperature fluctuations to mechanical anomalies—right on the edge. This real-time processing enables robots to adjust their actions without delay, thus enhancing efficiency and reducing downtime. In a setting where milliseconds can translate to significant production losses, edge analytics provides a competitive edge.

One of the most striking benefits of edge analytics is its ability to facilitate predictive maintenance. Instead of relying on scheduled maintenance or reactive troubleshooting, manufacturing plants can leverage real-time data analytics to predict potential failures before they occur. By analyzing patterns such as vibrations, temperature changes, or motor sounds, humanoid robots can identify signs of wear and tear. This proactive approach reduces unexpected halts and optimizes operational continuity, saving both time and costs.

Another advantage is security. Processing data locally minimizes the amount of sensitive information transmitted over networks. This feature is particularly valuable in industrial settings, where the stakes of data breaches can be high. With edge analytics, data is analyzed and acted upon locally, reducing the window of vulnerability.

However, the benefits of edge analytics aren't just limited to operational efficiency and security. The technology also fosters a more dynamic interaction between human workers and robots. For instance, humanoid robots equipped with edge analytics can better interpret and respond to human actions and commands. They can monitor the environment in real time to ensure the safety of their human colleagues, making the workspace safer and more collaborative. By processing data locally, robots can adjust their movements and behaviors instantaneously in response to human activities, ensuring harmonious co-working conditions.

Moreover, the critical need for latency reduction in industrial settings underscores the importance of edge analytics. The delayed response times inherent in cloud computing configurations can hinder the swift decision-making required on factory floors. Whether it's adjusting the speed of an assembly line or detecting anomalies in real-time, edge analytics ensures that such decisions are made within milliseconds, contributing to the seamless performance of complex manufacturing processes.

Consider the automotive sector, renowned for its stringent quality control and high-speed production lines. In this industry, even a fleeting delay in decision-making can disrupt the meticulous choreography of robotic and human labor. Edge analytics, by processing data locally, removes the latency that could hamper such a delicately balanced ecosystem. This allows humanoid robots to maintain the high standards expected in automotive manufacturing, ensuring that quality is never compromised by slower, cloud-dependent data processing.

Besides operational perks, edge analytics help in resource management. Factories amass enormous amounts of data daily, and sending all this information to a central cloud for processing can be both time-consuming and costly. Edge analytics alleviates this burden by handling significant volumes of data locally. This decentralized approach not only speeds up data processing but also decreases bandwidth usage and reduces costs associated with data storage.

The role of edge analytics is further amplified when we delve into the customization capabilities of humanoid robots. Manufacturing settings often require personalized solutions for differing production needs. Edge analytics enables robots to adapt swiftly to such unique demands. For instance, in assembly lines where product specifications can vary frequently, edge analytics allows robots to adjust their operations on-the-fly, maintaining high levels of customization without sacrificing speed or accuracy.

Nevertheless, the implementation of edge analytics is not without its challenges. Technical issues related to the integration of hardware and software, as well as initial setup costs, can be considerable. However, these investments are quickly justified by the long-term gains in operational efficiency, security, and cost reductions.

Edge analytics is not a stand-alone miracle; it thrives in an ecosystem where edge and cloud computing work synergistically. The cloud still plays a crucial role in overseeing large-scale analytics, long-term data storage, and delivering updates to edge devices. But for real-time, mission-critical decisions, edge analytics is indispensable.

Looking ahead, the role of edge analytics in real-time decision making is set to expand significantly. As more factories embrace AI-powered humanoid robots, the demand for real-time data processing will only grow. Innovations in edge computing hardware and software are likely to make edge analytics even more efficient, cost-effective, and easier to implement. Factories equipped with these advanced systems will lead the charge in efficiency, safety, and customization, setting new benchmarks for the industry.

In conclusion, edge analytics is transforming how humanoid robots operate in manufacturing settings, enabling real-time decision making that enhances productivity, security, and human-robot collaboration. As these technologies advance, they promise to unlock even greater potentials, making our factories smarter, faster, and more adaptive than ever before.

Chapter 22:
Customization and Adaptability of Humanoid Robots

The future of industrial automation hinges on the ability of AI-powered humanoid robots to be customized and adapt to the unique demands of various industries. Unlike their traditional counterparts, these robots boast modular designs that allow for seamless upgrades and component swaps, ensuring they can evolve alongside technological advancements. Tailoring robots to perform specific tasks in diverse settings—be it automotive plants or electronics assembly lines—amplifies their utility and efficiency. This adaptability not only boosts productivity but also opens new avenues for industry-specific applications, making humanoid robots indispensable assets in modern manufacturing environments.

Tailoring Robots to Specific Industries

The adaptability and customizability of humanoid robots are paramount when integrating them into various industrial sectors. Each industry presents its distinct challenges and requirements, necessitating a tailored approach to leverage robotics technology effectively. The immense versatility of AI-powered humanoid robots provides the flexibility to tweak and modify their functionalities to fit specific industry needs.

One of the foremost benefits of humanoid robots tailored to specific industries is the optimization of task-specific attributes. Take the automotive sector as a prime example. In this field, precision and consistency are crucial. Humanoid robots can be customized with advanced machine learning algorithms to accurately replicate complex human tasks like welding, assembly, and quality control. They can be programmed to handle delicate components with care or exert significant force where necessary. These robots thus enhance the production line's efficiency while reducing human error and fatigue.

In the pharmaceutical and medical industries, the focus shifts to sterility, precision, and safety. Humanoid robots in these settings are often designed with specialized materials resistant to microbial contamination, ensuring they meet rigorous hygiene standards. They are programmed to handle sensitive tasks such as compounding medicines or assisting in surgeries, where even minor errors can have significant repercussions. Customization allows for the integration of high-precision sensors and imaging technologies, enhancing the robots' ability to perform delicate procedures with unmatched consistency and accuracy.

Similarly, the electronics industry benefits from humanoid robots tailored to manage intricate tasks with minute precision. The assembly of small components like microchips demands a high level of dexterity and accuracy. Custom humanoid robots equipped with fine motor control and high-resolution cameras can execute these tasks with precision, significantly minimizing defects and increasing overall productivity.

The food and beverage industry is another sector where customized humanoid robots significantly contribute. Here, robots must adhere to strict hygiene standards and be capable of tasks ranging from sorting and packaging to food preparation. Customizations such as washable and antimicrobial surfaces, coupled with the ability to operate in various temperature conditions, make them invaluable assets. These robots can be programmed to handle different food items delicately, ensuring no damage occurs during processing.

In construction, humanoid robots tailored to the industry's demands help address labor shortages and safety issues. These robots can be equipped with heavy-duty materials and advanced sensory systems to assist with lifting, welding, and assembling components in construction environments. Customizable software enables them to adapt to various construction projects' dynamic and often unpredictable nature, enhancing efficiency and safety on-site.

The agriculture sector also reaps benefits from customized humanoid robots. Robots in this industry are often tailored for specific tasks such as planting, harvesting, and monitoring crop health. Custom sensors and AI algorithms allow these robots to analyze soil conditions, identify pests, and optimize crop yields. The adaptability of humanoid robots ensures they can be programmed for different types of crops and farming techniques, providing farmers with versatile tools to enhance productivity and sustainability.

Moreover, industrial environments such as warehousing and logistics benefit from humanoid robots that improve inventory management, order fulfillment, and transportation tasks. Custom sensors for identifying and handling diverse materials and packages are critical in these settings. Integrating AI for route optimization allows these robots to navigate complex warehouse layouts efficiently, reducing operational costs and delivery times.

The adaptation process extends beyond physical and functional customizations to include seamless integration with existing systems and processes within these industries. Humanoid robots must communicate effectively with other machinery and human workers. Therefore, their operating systems and software architectures are customized for compatibility and interoperability. This ensures the entire ecosystem can function cohesively, leveraging data and insights from various sources for optimized decision-making.

Training and ongoing support are also integral to customizing humanoid robots for specific industries. As industries evolve, so do their needs. Customizability ensures that humanoid robots can be updated with new software and hardware enhancements to keep pace with technological advancements and changing industry standards.

Economic considerations also play a crucial role in tailoring humanoid robots to different sectors. Custom solutions often entail higher initial investments but yield significant long-term benefits such as faster production times, higher-quality outputs, and reduced operational costs. Industries must conduct thorough cost-benefit analyses to determine the financial viability of investing in customized humanoid robotics.

Furthermore, tailoring humanoid robots to specific industries fosters innovation and continuous improvement. Industries can experiment with various customizations to discover new applications and efficiencies. This iterative process leads to enhanced performance, productivity, and breakthroughs that might not be possible with a one-size-fits-all approach.

In conclusion, the ability to customize and tailor humanoid robots to meet the specific demands of various industries is a game-changer in today's industrial landscape. Whether in automotive, medical, electronics, food and beverage, construction, or agriculture, customized humanoid robots are not only enhancing efficiency and precision but also paving the way for safer, more sustainable, and innovative industrial practices. By embracing the unique needs of each industry and leveraging the flexibility of humanoid robotics, we can unlock the full potential of AI-powered automation for a more productive and prosperous future.

Charlie Addison

Modularity and Upgradability

In the landscape of industrial automation, one of the most compelling features of AI-powered humanoid robots is their inherent modularity and upgradability. In an ever-evolving technological environment, the ability to adapt and evolve is invaluable. Modularity allows for components to be customized, added, or replaced with minimal disruption to the existing system. Upgradability ensures that as technology advances, the robots can remain at the cutting edge without the need for complete replacement.

Let's start with modularity. Imagine a humanoid robot in a car manufacturing plant. This robot might be equipped with various interchangeable tools and sensors, such as welding torches, painting nozzles, or assembly tools. The base robot remains the same, but its functionality can be altered as needed. If a new task is introduced into the production line, there's no need to bring in a new robot entirely; instead, you swap out the relevant modules. It's like building with LEGO bricks, where each piece can be added or removed depending on the desired structure.

This modular approach provides immense flexibility, not just in daily operations but also in long-term strategic planning. For instance, a company facing a shift in market demand can promptly adapt their production line by reconfiguring the existing robots rather than investing heavily in new machinery. This translates to significant cost savings and a faster response time to market changes.

Upgradability complements modularity perfectly. Consider the rapid pace at which AI and robotics technologies are advancing. New algorithms are developed, processing units become faster, and sensors improve in accuracy and sensitivity. Upgradability means you can integrate these advancements into existing robots. Imagine being able to upgrade a robot's processing unit to handle more complex tasks or improving its sensor suite to achieve higher precision. This capability

184

keeps the robot fleet perpetually state-of-the-art without incurring the massive expense of total replacement.

The benefits extend beyond mere cost savings. Upgradable systems can sustain a competitive edge by continually improving performance and features. When a company can ensure its humanoid robots are always using the latest technologies, it places itself in a robust position within the market. This adaptability can be a decisive advantage in sectors where innovation and efficiency are pivotal for success.

Furthermore, modularity and upgradability foster a sustainable approach to industrial automation. By extending the lifespan of the humanoid robots through upgrades and modular changes, you reduce waste and the environmental impact associated with discarding outdated machinery. This aligns well with the growing emphasis on sustainability in manufacturing, an aspect that will be explored further in subsequent chapters.

To make modularity and upgradability a reality, manufacturers design robots with standardized interfaces and components. This ensures that new modules or upgrades can be seamlessly integrated into the existing system. For instance, compatible docking points for tools and universal communication protocols facilitate easy swapping and upgrading. Moreover, these standardized systems encourage a competitive marketplace, where various manufacturers can create interchangeable parts and software, driving innovation and reducing costs due to market competition.

Another crucial aspect to consider is the role of software in modularity and upgradability. Just as hardware modules can be swapped out or upgraded, the software governing these robots can be updated to enhance functionality or incorporate new AI algorithms. Cloud-based updates allow for instantaneous improvements across a fleet of robots, ensuring that they all benefit from the latest developments in AI. This is particularly vital in industrial settings where downtime can lead to significant financial losses.

In conclusion, the modularity and upgradability of humanoid robots represent a paradigm shift in industrial automation. These features offer unparalleled flexibility, cost efficiency, and sustainability, allowing manufacturers to stay ahead of the curve in a rapidly evolving technological landscape. They also underscore a broader trend towards smarter, more adaptable, and more sustainable manufacturing practices. As we delve deeper into the customization and adaptability of humanoid robots, it becomes evident that these capabilities are not just advantageous but essential for the future of manufacturing.

Chapter 23:
Ensuring Sustainability in Robotics Manufacturing

In this pivotal chapter, we address the imperative of ensuring sustainability within the realm of robotics manufacturing. It's not just about integrating AI and humanoid robots into factory systems; it's about doing so in a way that minimizes environmental impact and maximizes resource efficiency. By adopting eco-friendly manufacturing practices, companies can significantly reduce their carbon footprint and conserve valuable resources. The longevity and recyclability of robots also play a crucial role in this sustainable vision. Designing robots with durable materials and components that can be easily disassembled and reused ensures that we aren't contributing to electronic waste. Moreover, implementing closed-loop systems and energy-efficient protocols can drive the robotics industry towards a greener future. The goal is clear: to harmonize technological advancement with environmental stewardship, ensuring that the future of manufacturing not only excels in innovation but also in sustainability.

Eco-Friendly Manufacturing Practices

In an era where environmental sustainability is increasingly becoming a priority for industries worldwide, the implementation of eco-friendly manufacturing practices is essential. Robotics manufacturing is no exception. Efforts to ensure that the production of AI-powered humanoid robots aligns with sustainable principles are not just beneficial for the environment but can also enhance the long-term viability and social acceptance of these innovations.

The transition to eco-friendly manufacturing begins with the selection of raw materials. Robotics companies are increasingly opting for recyclable and biodegradable materials. Traditional components may contain harmful substances that lead to long-term ecological damage. In contrast, eco-friendly materials minimize the industry's footprint while maintaining the integrity and efficiency of the robots. It's a step that requires careful consideration and significant investment but reaps long-term rewards.

Moreover, advancements in 3D printing technology have revolutionized the way robotic parts are produced. This additive manufacturing process allows for precise material use, vastly reducing waste compared to traditional subtractive manufacturing methods. By using only the necessary amount of material for each component, waste is minimized, creating a more sustainable production cycle.

Energy consumption is another critical factor. Manufacturing processes can be energy-intensive, and transitioning to renewable energy sources is crucial. Robotics factories are now integrating solar panels, wind turbines, and other renewable energy technologies to power their operations. This transition not only reduces the carbon footprint but also sets a positive example for other industries to follow.

Water usage is an oft-overlooked aspect of eco-friendly manufacturing practices. Robotics manufacturing typically involves processes that use significant amounts of water for cooling and cleaning. Implementing closed-loop water systems, which recycle and reuse water, can significantly reduce the demand on freshwater resources and minimize waste discharge into the environment.

Factory design and layout also play an essential role. Optimizing factory layouts to enhance energy efficiency involves designing workflows that reduce unnecessary movement of materials and components. This can cut down on transportation emissions and energy use, creating a more streamlined and sustainable production environment. Automated systems and AI can be employed to continually assess and improve these layouts, ensuring ongoing sustainability improvements.

In addition, companies are turning to green logistics. For instance, electric vehicles and drones are being used for internal logistics, reducing reliance on fossil fuels. This aligns with holistic sustainability goals and ensures that every stage of the manufacturing process contributes to eco-friendliness.

Waste management is another area where significant improvements can be made. Robotics manufacturing generates various types of waste, from metal scraps to electronic waste (e-waste). Establishing robust recycling programs and collaborating with recycling firms ensure that these materials don't end up in landfills. Instead, they can be reprocessed and reused, closing the loop on the manufacturing cycle.

Partnerships and collaborations play a crucial role. Robotics manufacturers are increasingly forming alliances with environmental agencies and sustainability-focused organizations. These partnerships facilitate the exchange of best practices and the development of innovative solutions that bridge the gap between high-tech manufacturing and environmental stewardship.

Beyond the factory floor, it's essential to consider the environmental impact of the entire supply chain. Many companies are now adopting green supply chain management practices. This involves working closely with suppliers to ensure that all materials are sourced sustainably and that suppliers adhere to eco-friendly practices. It's about fostering a culture of sustainability that extends beyond the manufacturing facility.

Robotics manufacturers also benefit from adopting a life cycle assessment (LCA) approach. LCA involves evaluating the environmental impact of a robot from the extraction of raw materials through to its end-of-life disposal. By understanding and mitigating the impacts at each stage, companies can create truly sustainable products. Designing robots with end-of-life recycling in mind ensures that every component can either be reused or recycled, preventing unnecessary waste.

Education and awareness among the workforce are pivotal for the success of eco-friendly practices. Training employees on sustainability initiatives and encouraging them to adopt environmentally friendly practices at work fosters a culture of environmental responsibility. It also drives home the point that sustainability is an integral part of the company's ethos.

Technological innovations hold the key to sustainable practices. The use of advanced AI and machine learning algorithms can improve energy efficiency by optimizing machine operations in real-time. Predictive maintenance, powered by AI, ensures machines operate at peak efficiency, reducing energy consumption and extending the lifespan of equipment, thereby reducing waste.

Furthermore, as the industry moves towards more sustainable practices, transparency and reporting become vital. Robotics manufacturers should adopt transparent reporting mechanisms to share their sustainability metrics and progress with stakeholders. This approach not only builds trust but also drives continuous improvement as companies strive to meet and exceed their sustainability goals.

Finally, government policies and incentives play a crucial role. Robotics companies must stay informed about and actively engage with policies that promote sustainable manufacturing. Incentives like tax breaks and grants for companies that achieve specific sustainability benchmarks serve as a significant motivation for companies to invest in eco-friendly practices.

In conclusion, implementing eco-friendly manufacturing practices in robotics goes beyond the mere adoption of green technologies. It requires a holistic approach that integrates sustainable principles into every facet of the manufacturing process, from material selection and energy use to waste management and supply chain logistics. As we continue to push the boundaries of what's possible with AI-powered humanoid robots, ensuring these innovations are produced sustainably is vital for our environment and future generations.

Longevity and Recyclability of Robots

Ensuring the sustainability of robotics manufacturing is more than just a laudable goal; it's an absolute necessity in a world increasingly focused on environmental responsibility. The longevity and recyclability of robots are critical components of this endeavor, helping to reduce waste and conserve resources. By addressing these aspects, companies can create robots that not only perform excellently but also contribute positively to our planet's well-being.

To begin with, the concept of longevity in robots implies designing machines that have extended operational lifespans. This involves the use of high-quality, durable materials that can withstand the rigors of industrial environments. Robust construction and the capability to function effectively even under continuous operation are crucial. Companies are now focusing on integrating advanced diagnostic tools that predict potential failures before they occur. By employing predictive maintenance strategies, downtime is minimized, and the robots can achieve consistent performance over many years.

The shift towards modular design plays a pivotal role in enhancing the longevity of humanoid robots. Modular components can be replaced or upgraded individually without necessitating a complete overhaul of the robot. This not only extends the functional life of the robot but also allows for continual improvements as newer technologies become available. Modular systems mean that outdated parts can be swapped out for the latest hardware, ensuring the robot remains state-of-the-art even many years after its initial deployment.

Another significant aspect is the software longevity. Sophisticated AI-driven robots require regular software updates to enhance their capabilities and security. Unlike hardware, which has physical wear and tear, software can be continually improved without being constrained by physical degradation. Consequently, a well-designed software architecture that supports easy updates can ensure that a robot remains relevant and efficient throughout its operational life.

In terms of recyclability, it's imperative that robotics manufacturers adopt a cradle-to-cradle philosophy. This approach focuses on designing robots with end-of-life disposal in mind. By selecting materials that can be easily dismantled and recycled, companies can significantly reduce the environmental impact of discarded robots. For instance, metals like aluminum and steel commonly used in robot frames are highly recyclable. Plastics and other composite materials, while more challenging, can also be designed for easier recycling.

The disassembly process must also be streamlined for recyclability. Robots conceived with disassembly in mind enable more efficient recycling operations. If components are easy to detach and categorize, the recycling process becomes less labor-intensive, less costly, and significantly more effective. This design philosophy encourages manufacturers to consider the entire lifecycle of the robot, from raw material extraction to end-of-life disposal.

Moreover, incorporating eco-friendly materials into the manufacturing process is another step forward. Biodegradable materials, recycled plastics, and non-toxic chemicals are becoming increasingly popular in the creation of various robot parts. While there's still a long way to go in perfecting these materials for industrial application, the progress thus far is promising. As research in this area continues to advance, the hope is to achieve a manufacturing process that's not only efficient but also sustainable in every respect.

Corporate responsibility goes beyond just the production phase. Companies should also offer programs to take back and recycle their old robots. By facilitating the return of outdated or non-functional robots, manufacturers can ensure these machines are disposed of in an environmentally friendly manner. This practice not only aids in resource conservation but also fosters customer loyalty, as businesses see manufacturers taking responsibility for their products throughout the entire lifecycle.

In addition to material considerations, energy efficiency during the robot's operational life is crucial. Robots that consume less energy not only reduce operational costs but also lessen the environmental footprint. Energy-efficient motors and power management systems are essential components. Companies are investing in cutting-edge innovations such as regenerative braking systems, which capture and reuse energy that would otherwise be lost. Efficient energy use extends the operational time of robots and reduces the frequency of battery replacements, contributing to both longevity and sustainability.

Finally, the concept of reuse should not be overlooked. Although recycling old robots into their base materials is beneficial, reusing entire systems or subsystems can be even more effective. For example, robots that have outlived their initial purpose can be refurbished and repurposed. This practice not only conserves resources but also provides a cost-effective solution for companies that might not afford brand-new robots. By creating a market for used robots, we can ensure these machines are fully utilized before they reach the recycling phase.

As we look to the future, the longevity and recyclability of robots will likely play central roles in the robotics industry. Innovations in materials science, software engineering, and manufacturing techniques will drive this change. The goal will be to create a closed-loop system where robots are produced, used to their fullest extent, and then efficiently recycled or repurposed, thereby achieving true sustainability.

Given the growing attention to climate change and resource scarcity, it's clear that the robotics industry must evolve. By prioritizing longevity and recyclability, we can ensure that robots not only advance industrial capabilities but also contribute to a healthier, more sustainable planet. The journey toward sustainable robotics manufacturing is complex and challenging, but with continued effort and innovation, it is undoubtedly achievable. Ultimately, the commitment to these principles will define the success of robotics in modern industrial settings, balancing technological advancement with ecological responsibility.

Chapter 24:
The Road Ahead: Preparing for Widespread Adoption

The path to widespread adoption of AI-powered humanoid robots in industrial settings is both thrilling and challenging. As we stand on the brink of this technological revolution, industries must carefully navigate the hurdles ahead. Key factors such as cost constraints, workforce adaptation, and regulatory landscapes play pivotal roles in shaping the future. Companies need comprehensive roadmaps to seamlessly integrate these sophisticated machines without disrupting existing workflows. By fostering a culture of continuous innovation and collaboration, industries can maximize the benefits of humanoid robots, driving unparalleled efficiency and creativity. Preparing for this transformation involves not just technological readiness but also a commitment to upskilling the current workforce and cultivating public trust. As we forge ahead, the collective effort of stakeholders will determine how successfully we embrace these advancements, ultimately setting the stage for an era where human ingenuity and robotic precision coalesce to redefine manufacturing paradigms.

Overcoming Barriers to Entry

As we navigate the path toward widespread adoption of AI-powered humanoid robots in industrial settings, it's crucial to address the barriers to entry that currently impede progress. While the advantages of integrating these advanced robots are numerous, ranging from increased efficiency to reduced errors, several significant challenges need to be overcome. In this section, we'll explore these obstacles and discuss potential strategies to mitigate them, paving the way for a smoother transition to robot-enhanced manufacturing environments.

The first major barrier is the high initial investment required for purchasing and implementing advanced humanoid robots. These robots are still relatively expensive, primarily due to the sophisticated AI and machine learning technologies they incorporate, as well as the cutting-edge sensor and material components they require. This financial hurdle makes it difficult for small and medium-sized enterprises (SMEs) to justify the costs, especially when their profit margins are already thin. To address this issue, companies could explore financing options such as leasing or subscription models, which spread the costs over time and reduce the immediate financial burden. Governments and industry organizations could also offer grants or incentives to encourage adoption among SMEs.

Another significant barrier is the lack of skilled personnel capable of maintaining and operating these advanced robots. Unlike traditional manufacturing machinery, AI-powered humanoid robots require expertise in programming, AI, and robotics. The current workforce may lack these specialized skills, which can hinder the effective deployment and utilization of these robots. To overcome this obstacle, educational institutions and industry leaders must collaborate to develop comprehensive training programs and curricula focused on robotics and AI. Apprenticeships and on-the-job training can also be valuable avenues for building the necessary skill sets within the existing workforce.

Resistance to change is an often-overlooked barrier but one that can significantly impact the adoption process. Employees and managers alike may be wary of new technologies that disrupt established workflows and job roles. This resistance can stem from a fear of job displacement, frustration with learning new systems, or simply a preference for familiar methods. Addressing these concerns requires clear communication about the benefits of humanoid robots, involving employees in the transition process, and demonstrating how these technologies can augment rather than replace human labor. Providing reassurances about job security and offering reskilling programs can also help alleviate anxieties.

Interoperability with existing systems is another technical hurdle that companies face when integrating humanoid robots. Industrial environments are often characterized by a mix of legacy machinery and proprietary software, creating compatibility issues when new technologies are introduced. Ensuring seamless integration with existing systems is critical to maximizing the benefits of AI-powered robots. Industry standards and open-source software platforms can play a crucial role in facilitating this interoperability. Collaborative efforts among tech developers, manufacturers, and standards organizations are essential to develop universally accepted protocols and frameworks.

Furthermore, data security and privacy concerns present another formidable barrier. As humanoid robots become more integrated into manufacturing processes, they generate and rely on vast amounts of data, which often include sensitive proprietary information. Ensuring the security of this data and protecting it from cyber threats is paramount. Implementing robust cybersecurity measures, such as encryption, firewalls, and regular security audits, is indispensable. Additionally, adhering to best practices and regulatory requirements related to data privacy will help build trust among stakeholders.

Scalability is an additional consideration that must be addressed for widespread adoption. While initial pilot programs and small-scale implementations can provide valuable insights and proof of concept, scaling up these initiatives to cover entire production lines or multiple factories poses its own set of challenges. These include ensuring consistent performance across different environments, managing increased data loads, and maintaining reliability. Developing modular systems and scalable architectures can help mitigate these issues, allowing companies to expand their use of humanoid robots incrementally and with greater control.

On the technical front, overcoming software and hardware limitations remains a significant challenge. AI-powered humanoid robots rely on advanced algorithms and high-performance hardware to function effectively. However, continuous advancements in AI, machine learning, and robotics hardware are essential to keep up with the evolving demands of industrial applications. Ongoing research and development, coupled with collaborative efforts between academia and industry, will be key to pushing the boundaries of what these robots can achieve.

Addressing regulatory and compliance hurdles is also crucial. Different countries and regions have varying regulations related to the deployment of advanced robotics and AI technologies in industrial settings. These regulations often encompass safety standards, ethical considerations, and operational guidelines. Navigating this complex landscape requires companies to stay informed about relevant regulations and actively engage with policymakers to ensure that their robots comply with local laws. Establishing industry-wide standards and best practices can help create a more unified regulatory environment, facilitating smoother adoption across different regions.

Finally, fostering a culture of innovation and continuous improvement within organizations is vital for the successful adoption of humanoid robots. Companies must cultivate an environment where experimentation and innovation are encouraged and celebrated. This cultural shift can help overcome resistance to change and enable organizations to adapt more readily to new technologies. Encouraging cross-functional collaboration, promoting open communication, and recognizing and rewarding innovative initiatives can contribute to building this culture.

In summary, while there are several barriers to entry that must be addressed for the widespread adoption of AI-powered humanoid robots in industrial settings, none of these challenges are insurmountable. By focusing on reducing initial costs, developing skilled personnel, fostering a culture of innovation, and addressing technical and regulatory hurdles, companies can pave the way for successful integration of these advanced technologies. Overcoming these barriers will not only enhance manufacturing efficiency and productivity but also position industries to thrive in a rapidly evolving technological landscape.

Building a Roadmap for Integration

Preparing for the widespread adoption of AI-powered humanoid robots in industrial settings is no small feat. It requires a meticulously planned roadmap to navigate the complexities and ensure seamless integration. First, it's crucial to identify the industries where humanoid robots can deliver the most substantial benefits. Industries with repetitive, hazardous, or highly intricate tasks stand to gain immensely. In the beginning, operational assessments and feasibility studies should be conducted to ascertain where these robots can be most impactful.

A roadmap for integration should begin with pilot projects. These projects serve as trial runs that help to identify potential issues before a full-scale deployment. A well-structured pilot project allows for testing the robots in real-world conditions, evaluating both their performance and their interaction with human workers. Here, it's essential to gather extensive data and feedback to make informed decisions and adjustments. Pilot projects provide invaluable insights that can highlight unforeseen challenges and opportunities for improvement.

Next, the integration process should include a detailed timeline that outlines the various stages of implementation. This timeline serves as a visual guide that helps stakeholders understand the sequence of events and the dependencies between different tasks. A phased approach is often most effective, allowing companies to gradually scale up their use of humanoid robots. This minimizes disruption and promotes smoother transitions, enabling both the technology and human workers to adapt progressively.

Training programs for human workers should coincide with the technological rollout. These training initiatives should be comprehensive, covering not only the operational aspects of working alongside humanoid robots but also addressing any concerns employees might have. By involving staff early in the process, companies can ensure a smoother transition and foster a collaborative environment. Training should emphasize the value of human-robot collaboration, mitigating fears of job displacement and highlighting opportunities for skill enhancement.

During the early stages of integration, close monitoring and constant evaluation are imperative. Keeping a pulse on the robot's performance and its interaction with human workers allows for real-time troubleshooting and optimization. This stage is critical for gathering performance data, which can be used to tweak algorithms, update software, and refine operational protocols. Continuous improvement loops should be established to ensure that any issues are promptly addressed, and the robots' functionalities are continuously enhanced.

Budgeting is another key aspect of the integration roadmap. Funding should be allocated not just for the initial purchase and deployment of the robots but also for ongoing maintenance, updates, and staff training. Financial planning must take into account potential cost overruns and provide contingencies for unforeseen challenges.

Investing in robust support structures, such as a dedicated technical team, ensures that the robots remain operational and efficient in the long run.

To facilitate seamless integration, companies must also focus on developing or adopting standardized interfaces and communication protocols. These standards ensure compatibility with existing manufacturing systems and infrastructure. Interoperability is key to integrating humanoid robots without the need for extensive reengineering of current processes. Open standards also enable scalability, allowing new robots to be easily added to the system as operations expand.

Collaboration with technology providers and industry experts can significantly ease the integration process. Partnerships with established robotics firms bring in-depth technical know-how and support. Engaging with academic institutions can also provide cutting-edge research insights and innovative solutions. By leveraging external expertise, companies can enhance their integration efforts and stay ahead of technological advancements.

Regulatory compliance is another critical factor. Adhering to industrial safety standards and obtaining the necessary certifications is essential to avoid legal pitfalls and ensure the safety of both human workers and robots. Companies should stay updated on evolving regulations and actively participate in industry forums to help shape future standards. Being proactive in compliance not only safeguards operations but also builds trust with employees and stakeholders.

Additionally, effective communication strategies should be developed to manage internal and external perceptions. Clearly articulating the benefits of humanoid robots and addressing any fears or misconceptions can foster a positive attitude towards the integration process. Regular updates, open forums, and transparent discussions create a culture of trust and acceptance.

The integration roadmap must also account for the long-term evolution of both the technology and the industry. Building a flexible and adaptable strategy allows companies to pivot as new innovations emerge. Future-proofing the integration plan ensures it remains relevant and effective despite rapid technological advancements. This forward-thinking approach prepares companies to continuously optimize their use of humanoid robots and sustain competitive advantage.

Lastly, fostering a culture of innovation within the organization is essential for the successful adoption of humanoid robots. Encouraging employees to engage in continuous learning and to contribute ideas can drive the ongoing optimization of robotic systems. By viewing the integration process as a collaborative effort, companies can unleash collective creativity and problem-solving capabilities.

In essence, building a roadmap for integration is a multi-faceted endeavor that demands thorough planning, education, and collaboration. From pilot projects and phased rollouts to training programs and regulatory compliance, each step must be meticulously managed. With a well-crafted roadmap, industries can unlock the full potential of AI-powered humanoid robots, revolutionizing production processes and paving the way for a transformative future in manufacturing.

Chapter 25:
Case Study A Day in the Life of a
Factory with Humanoid Robots

In the heart of a bustling factory, the seamless integration of humanoid robots with human workers is transforming the landscape of industrial production. As the morning shift begins, robot assistants are already hard at work, handling repetitive tasks with precision that's unmatched, freeing human workers to focus on more complex problem-solving activities. These robots, equipped with advanced AI and machine learning algorithms, continually adapt to the dynamic environment, improving their efficiency over time. As the day progresses, the harmonious collaboration between humans and robots becomes even more evident; robots assist in lifting heavy materials, ensuring accuracy in assembly lines, and even performing quality checks with impeccable precision. By the end of the shift, the factory has not only met its production targets but has also witnessed a significant reduction in errors and downtime. This case study vividly illustrates the immense benefits and transformative potential of humanoid robots, not just as tools but as partners in the pursuit of industrial excellence.

Detailed Walkthrough of Daily Operations

In this section, we dive into the daily operations of a factory that has fully embraced humanoid robots as part of its workforce. The goal is to provide an inside look at how these advanced machines integrate into the complex environment of industrial manufacturing, working seamlessly alongside human employees to enhance productivity and efficiency.

At the break of dawn, the factory comes to life with the quiet but efficient activity of humanoid robots preparing for the day's tasks. These robots, powered by advanced AI algorithms and machine learning models, begin their day by syncing with the central control system, ensuring they have the latest updates on production schedules, inventory levels, and machine statuses. This initial synchronization is crucial as it allows the robots to plan their activities and coordinate with one another effectively.

The factory floor is a meticulously organized space divided into several zones, each designated for specific parts of the manufacturing process. In Zone 1, humanoid robots specialize in handling raw materials. They deftly unload trucks, using their robust sensors and data interpretation capabilities to sort and store materials accurately. Their ability to lift heavy loads with precision and without fatigue ensures that the raw materials are ready for the next steps in the production line.

Meanwhile, in Zone 2, robots equipped with sophisticated vision systems and precision tools take on the task of assembling components. These robots work in concert, communicating wirelessly to pass parts to each other in a seamless ballet of mechanical efficiency. Their built-in error detection systems constantly monitor for any anomalies, ensuring that any defects are immediately identified and corrected. This not only maintains high standards of quality but also minimizes waste and rework.

Human workers in the factory are not sidelined; instead, they take on supervisory and strategic roles. In Zone 3, operators oversee the work of humanoid robots, ensuring everything runs smoothly. These workers focus on tasks that require human judgment and creativity, such as problem-solving and optimization of production processes. The collaboration between humans and robots exemplifies the synergy of combining human ingenuity with robotic precision.

As the day progresses, the robots transition through different zones, each with a specialized function. In Zone 4, they handle the task of packaging finished products. Humanoid robots equipped with advanced gripping technologies and vision systems ensure that products are packaged efficiently and correctly. Their ability to adapt to different product sizes and packaging requirements adds a layer of flexibility previously unattainable in traditional production lines.

Robotic maintenance crews consist of specialized humanoid robots assigned to monitor and repair factory equipment. These robots are equipped with diagnostic tools and AI-driven predictive maintenance algorithms that can foresee equipment failures before they occur. Regular inspections and timely interventions not only reduce downtime but also extend the lifespan of the machinery, contributing to overall operational efficiency.

Another remarkable aspect of daily operations is the role of robots in inventory management. In Zone 5, inventory robots equipped with RFID scanners and autonomous navigation systems move through the warehouse, tracking inventory levels and ensuring precise record-keeping. Their ability to swiftly locate items and update the central database in real-time aids in maintaining optimal inventory levels, thus preventing overstocking or shortages.

Lunch breaks are staggered not just for human employees but also for humanoid robots. During these intervals, robots enter standby modes where they perform self-diagnostics and recharge their batteries. This period of downtime ensures that they are operating at peak efficiency and are ready to resume their tasks without interruption.

Communication is key in such a dynamic environment. Throughout the day, humanoid robots constantly exchange data with each other and the central control system. This real-time communication allows for quick adjustments to be made to the production schedule in response to any changes or unforeseen issues. For example, if a particular component is running low, inventory robots can swiftly alert the supply chain management team, ensuring timely replenishment.

Afternoon shifts see an increase in collaboration between human workers and humanoid robots in product testing and quality assurance in Zone 6. Robots conduct a battery of tests, using their sensors and data-analysis capabilities to verify the integrity of each product. Human workers review and validate the findings, making decisions on any necessary adjustments or recalls.

The end of the day involves a detailed data review process. Robots upload logs and performance metrics to the central system, which analyzes the data to provide insights into production efficiency, error rates, and areas for improvement. Human supervisors review these insights, using them to plan for the next day's operations and make strategic decisions to optimize the manufacturing process further.

In conclusion, a day in the life of a factory with humanoid robots is a testament to the power of advanced technology in transforming industrial operations. By taking on repetitive and physically demanding tasks, these robots allow human workers to focus on higher-level functions, resulting in a harmonious and highly productive environment. The seamless integration of humanoid robots into daily operations not only enhances efficiency but also paves the way for innovative production methods and a new era in manufacturing.

Real-World Benefits and Insights

In the midst of our exploration into the factory floor bustling with humanoid robots, it's crucial to extract and analyze the tangible benefits these robots have brought to industrial environments. While the idea of humanoid robots might conjure images of science fiction, their real-world applications are far-reaching and profoundly impactful.

One of the foremost benefits noticeable in a factory equipped with humanoid robots is the significant uptick in efficiency. Unlike traditional industrial robots, AI-powered humanoids possess the dexterity and intelligence to perform a multitude of tasks without human intervention. They are programmed to learn and adapt, allowing them to handle complex, repetitive tasks with precision and speed. This results in a dramatic reduction in production times, helping companies meet ever-increasing consumer demands.

Moreover, these robots bring about a notable improvement in quality. In an era where precision is paramount, AI and machine learning algorithms enable humanoid robots to detect defects and inconsistencies that might elude human eyes. By ensuring that every product meets stringent quality standards, these robots minimize errors and rework, ultimately leading to a more reliable and consistent output. This is especially valuable in industries such as automotive and electronics, where even minor defects can lead to significant issues down the line.

The integration of humanoid robots into factory environments also enhances safety. Unlike their human counterparts, robots aren't susceptible to fatigue or distraction, two common factors contributing to workplace accidents. These robots can be assigned hazardous or physically demanding tasks, thereby reducing the risk of injuries to human workers. In sectors where exposure to harmful substances or extreme conditions is inevitable, the presence of robots ensures that human workers are shielded from potential harm.

Additionally, the implementation of AI-powered robots fosters a transformative shift in workforce dynamics. Contrary to the fear that robots might replace human jobs, they are creating opportunities for upskilling and career advancement. Workers are transitioning from performing manual, repetitive tasks to taking on roles that require a higher level of cognitive skill and problem-solving capabilities. Companies are investing in training programs to help their employees adapt to this new landscape, paving the way for a more knowledgeable and versatile workforce.

Another insightful benefit of utilizing humanoid robots in factories is the enhancement of data analytics and decision-making processes. These robots generate vast amounts of data through their sensors and operational activities. By leveraging big data analytics, companies can gain actionable insights into their production processes, identify bottlenecks, and forecast maintenance needs. This predictive maintenance capability helps in preventing costly downtime, ensuring that machinery is always in optimal condition.

On the scalability front, humanoid robots offer unmatched flexibility. Businesses can easily scale their operations up or down by deploying additional robots or reallocating existing ones to different tasks as needed. This level of adaptability is vital in today's fast-paced market, where demand fluctuations are common. Whether a factory needs to ramp up production for a seasonal surge or adjust to a sudden drop in orders, humanoid robots provide the necessary agility to respond effectively.

The environmental impact of humanoid robots is another area where significant benefits can be seen. With their ability to optimize production processes and reduce waste, these robots contribute to more sustainable manufacturing practices. They can operate using algorithms that prioritize energy efficiency, thereby lowering the factory's overall carbon footprint. In an age where sustainability is a critical concern, integrating robotics into manufacturing not only supports corporate environmental goals but also appeals to eco-conscious consumers.

Financially, the long-term cost savings stemming from the use of humanoid robots are substantial. While the initial investment might be high, the return on investment (ROI) manifests through reduced labor costs, decreased error-related expenses, and lower maintenance expenditures due to proactive troubleshooting. Companies can achieve a more streamlined, cost-effective operation that enhances their competitive edge in the global market.

In terms of innovations and technological advancements, humanoid robots are catalysts for continuous improvement. Their integration into factory settings drives R&D initiatives, leading to the development of more advanced and capable robotic systems. This cycle of innovation and improvement results in a dynamic manufacturing environment that is constantly evolving to meet new challenges and opportunities.

Overall, the deployment of AI-powered humanoid robots is not just about replacing human labor; it's about augmenting human capabilities and creating a collaborative ecosystem. This synergistic approach fosters a workspace where humans and robots complement each other's strengths, leading to unprecedented levels of productivity and innovation. As factories around the world continue to embrace this technology, we can expect to see even more profound transformations in the industrial sector.

In conclusion, the real-world benefits of humanoid robots extend across efficiency, quality, safety, workforce dynamics, data-driven decision-making, scalability, environmental sustainability, financial savings, and innovation. These insights underscore the potential for

AI-powered humanoid robots to revolutionize factory operations, offering a glimpse into a future where technology and humanity work hand in hand to create a more efficient, safe, and prosperous industrial landscape.

Embracing the Future of AI and Robotics in Manufacturing

The journey through the chapters of this book has provided a comprehensive insight into how AI-powered humanoid robots are poised to revolutionize the manufacturing world. As we look to the future, it's clear that the marriage of artificial intelligence and robotics will significantly impact every aspect of industrial production, from efficiency and safety to economic and societal dimensions. The transformation promises not only technical advancements but also a new paradigm in the way humans and machines collaborate.

The striking evolution of humanoid robots, driven by sophisticated AI and machine learning algorithms, represents a quantum leap from traditional automation. This technological leap is already beginning to transform manufacturing processes, enabling unprecedented levels of precision and efficiency. But beyond these immediate benefits, these advancements hint at something far more significant: a fundamental shift in how we conceptualize work and productivity. As manufacturers continue to adopt and integrate these innovations, they're setting the stage for a new industrial era.

Looking ahead, the integration of AI-powered humanoid robots will require a careful balance between leveraging technology and preserving human roles. The optimization of human-robot collaboration is paramount, necessitating ongoing training and a reevaluation of human skills. Factories of the future will be dynamic environments where robots and humans learn from each other, increasing operational efficiency while ensuring that human workers are engaged in more meaningful and less hazardous tasks.

One cannot ignore the economic implications of widespread AI and robotic integration. On one hand, companies stand to benefit from significant cost savings and productivity gains. On the other, they face the challenge of workforce restructuring as certain jobs become obsolete while new roles emerge. Sectors need to prepare for this transition by investing in retraining programs that align human skills with the demands of a robot-centric workplace.

Equally significant are the ethical and societal implications. The deployment of humanoid robots in the workforce raises questions about job displacement, income inequality, and the need for new regulatory frameworks. Public perception and acceptance will be critical, and ethical considerations must guide the development and integration of these technologies. As we design and implement these robotic systems, fostering a transparent dialogue around these issues will be crucial to ensuring a balanced and socially responsible advancement.

The lessons learned from early adopters and leading companies provide a valuable roadmap for those still on the cusp of embracing AI-powered robotics. Case studies illustrate the tangible benefits, from increased production speed and quality to improved safety and operational efficiency. These success stories also highlight the importance of adaptability and continuous learning in harnessing the full potential of humanoid robots.

Future trends point to an even deeper integration of these technologies. Innovations in sensor technology, real-time data processing, and advanced algorithms will continue to enhance robotic capabilities. Humanoid robots are becoming more adept at complex tasks and interacting seamlessly with human counterparts. Emerging industries, from pharmaceuticals to aerospace, will see unique applications of these technologies, expanding the horizons of what's possible in manufacturing.

Sustainability is an essential aspect of future manufacturing, and AI-powered robotics plays a crucial role in this context. By optimizing resource use, reducing waste, and enabling predictive maintenance, these technologies contribute to more eco-friendly production practices. Moreover, the longevity and recyclability of robots themselves will be critical in minimizing their environmental impact.

As we prepare for widespread adoption of AI and robotics, it's crucial to overcome several barriers to entry. This includes addressing technical challenges, ensuring regulatory compliance, and fostering public acceptance. Building a strategic roadmap that outlines clear steps for integration will be essential for companies aiming to stay competitive in this rapidly evolving landscape.

The future of AI and robotics in manufacturing is not just about technological advancements but also about fostering a culture of innovation and continuous improvement. By leveraging big data and analytics, manufacturers can drive better decision-making, optimize operations, and unlock new opportunities for growth. Cloud computing and edge AI further enhance these capabilities, providing robust solutions for real-time processing and decision-making.

The adaptability and customization of humanoid robots will be key in tailoring solutions to specific industrial needs. Modularity and upgradability ensure that these robotic systems remain relevant and effective as technology evolves. This flexibility will be a significant advantage in meeting the diverse and changing demands of various manufacturing sectors.

In conclusion, embracing the future of AI and robotics in manufacturing requires a holistic approach. It involves not just adopting new technologies but also rethinking our approaches to work, training, ethics, and sustainability. By doing so, we can harness the full potential of these advancements to create more efficient, safe, and socially responsible manufacturing environments.

The road ahead is both challenging and exciting. As we stand on the brink of this new era, the possibilities are boundless. The integration of AI-powered humanoid robots in manufacturing promises to reshape industries, drive economic growth, and improve the quality of life for future generations. By embracing these changes with a thoughtful and proactive approach, we open the door to a more prosperous and innovative future.

Glossary of Terms and Acronyms

This glossary provides definitions for key terms and acronyms used throughout the book. It serves as a quick reference for readers to understand the specialized vocabulary related to industrial automation and AI-powered humanoid robots.

A

AI (Artificial Intelligence): The simulation of human intelligence processes by machines, especially computer systems.

Analytics: The systematic computational analysis of data or statistics to discover, interpret, and communicate meaningful patterns.

B

Big Data: Extremely large data sets that may be analyzed computationally to reveal patterns, trends, and associations.

C

Collaborative Robots (Cobots): Robots designed to work alongside humans in a shared workspace.

Cloud Computing: The delivery of computing services over the internet, allowing for flexible resources and scalability.

D

Data Interpretation: The process of reviewing data with the aim to draw conclusions from it.

E

Edge AI: The use of artificial intelligence algorithms in edge computing devices to process data locally, closer to where it is generated.

Ergonomics: The study of people's efficiency in their working environment, aiming to improve interaction between humans and machines.

F

Factory Automation: The use of control systems, such as computers or robots, for handling different processes and machinery in an industry to replace human intervention.

H

Humanoid Robots: Robots designed to resemble and mimic human body movements and interactions.

I

Industrial Robotics: Automated, programmable, and capable of movement on three or more axes, used in manufacturing and related sectors.

Intelligent Robotics: Robots enhanced with AI to perform complex tasks by perceiving their environment and making decisions autonomously.

IoT (Internet of Things): The network of physical devices that are embedded with sensors, software, and other technologies to connect and exchange data with other devices and systems over the internet.

M

Machine Learning: A subset of AI that involves the use of algorithms and statistical models that enable computers to perform tasks without explicit instructions.

Modularity: The degree to which a system's components may be separated and recombined, often to allow for flexibility and variety in use.

P

Predictive Maintenance: Techniques that use data analysis tools and techniques to detect anomalies in your operation and possible defects in equipment and processes so you can fix them before they result in failure.

R

Real-Time Data Processing: The immediate processing of data upon collection, allowing for instant outputs and decisions.

Robotics: The branch of technology that deals with the design, construction, operation, and application of robots.

S

Sensor Technologies: Devices or systems that detect changes in the environment and send information to other electronics, often used in robots to perceive their surroundings.

T

Training: The process of teaching a machine or human to perform tasks by providing examples, instructions, or rules.

Traditional Robots: Robots that are pre-programmed to perform specific tasks with limited flexibility or adaptability.

www.ingramcontent.com/pod-product-compliance
Lightning Source LLC
Chambersburg PA
CBHW051233050326
40689CB00007B/911